Silicon Reagents
in Organic Synthesis

BEST SYNTHETIC METHODS

Series Editors

A. R. Katritzky
University of Florida
Gainesville, Florida
USA

O. Meth-Cohn
Sterling Organics Ltd
Newcastle on Tyne
UK

C. W. Rees
Imperial College of Science
and Technology
London, UK

Richard F. Heck, *Palladium Reagents in Organic Syntheses,* 1985
Alan H. Haines, *Methods for the Oxidation of Organic Compounds: Alkanes, Alkenes, Alkynes, and Arenes,* 1985
Paul N. Rylander, *Hydrogenation Methods,* 1985
Ernest W. Colvin, *Silicon Reagents in Organic Synthesis,* 1988

In preparation
Andrew Pelter, Keith Smith and Herbert C. Brown, *Borane Reagents,* 1988
Basil Wakefield, *Organolithium Methods,* 1988

Silicon Reagents
in Organic Synthesis

Ernest W. Colvin

Chemistry Department
University of Glasgow
Glasgow G12 8QQ
UK

1988

Academic Press
Harcourt Brace Jovanovich, Publishers
London   San Diego   New York   Berkeley
Boston   Sydney   Tokyo   Toronto

ACADEMIC PRESS LIMITED
24–28 Oval Road
London NW1 7DX

US Edition published by
ACADEMIC PRESS INC.
San Diego, CA 92101

This book is a guide providing general information concerning its subject matter; it is not
a procedural manual. Synthesis of chemicals is a rapidly changing field. The reader should
consult current procedural manuals for state-of-the-art instruction and applicable govern-
ment safety regulations. The Publisher and the authors do not accept responsibility for
any misuse of this book, including its use as a procedural manual or as a source of specific
instructions

British Library Cataloguing in Publication Data
Colvin, Ernest W.
   Silicon reagents in organic synthesis.—
   (Best synthetic methods)
   1. Chemistry, Organic—Synthesis
   2. Organosilicon compounds
   3. Chemical tests and reagents
   I. Title II. Series
   547'.2   QD262

   ISBN 0-12-182560-4

Typeset in Great Britain by
EJS Chemical Composition, Midsomer Norton, Bath BA3 4BQ

Printed in Great Britain by
St Edmundsbury Press Ltd, Bury St Edmunds, Suffolk

Contents

Foreword

There is a vast and often bewildering array of synthetic methods and reagents available to organic chemists today. Many chemists have their own favoured methods, old and new, for standard transformations, and these can vary considerably from one laboratory to another. New and unfamiliar methods may well allow a particular synthetic step to be done more readily and in higher yield, but there is always some energy barrier associated with their use for the first time. Furthermore, the very wealth of possibilities creates an information-retrieval problem. How can we choose between all the alternatives, and what are their real advantages and limitations? Where can we find the precise experimental details, so often taken for granted by the experts? There is therefore a constant demand for books on synthetic methods, especially the more practical ones like *Organic Syntheses*, *Organic Reactions*, and *Reagents for Organic Synthesis*, which are found in most chemistry laboratories. We are convinced that there is a further need, still largely unfulfilled, for a uniform series of books, each dealing consisely with a particular topic from a *practical* point of view—a need, that is, for books full of preparations, practical hints and detailed examples, all critically assessed, and giving just the information needed to smooth our way painlessly into the unfamiliar territory. Such books would obviously be a great help to research students as well as to established organic chemists.

We have been very fortunate with the highly experienced and expert organic chemists who, agreeing with our objective, have written the first group of volumes in this series, *Best Synthetic Methods*. We shall always be pleased to receive comments from readers and suggestions for future volumes.

A.R.K., O.M.-C., C.W.R.

vii

Preface

The past two decades have seen a remarkable growth in the involvement of silicon in modern organic chemistry; this growth to a large extent reflects the painstaking researches of the early pioneers of organosilicon chemistry, an area initially of little promise and considerable intractibility. It is intended that this book will provide a relatively painless introduction to practical organosilicon chemistry; the emphasis throughout is on the ability of silicon to act as a "ferryman". Silicon is normally absent from the product, but its temporary presence has selectively directed the course of the particular reaction or transformation under discussion.

The basic aim of the book is twofold: it attempts to describe routes to functionalized organosilanes, and, equally importantly, to describe the useful behaviour of such silanes, with constant emphasis being placed on the organic moiety. One thing it does not attempt to do is to provide a full coverage of organosilicon chemistry; pertinent references to more detailed treatments are given at appropriate points in the text, and relevant reviews are listed in Chapter 2. In general, only those processes/sequences for which full experimental accounts have been published are discussed in detail: this means that preliminary (and often the original) communications may not be cited directly.

From the outset, it has been assumed that the reader/practitioner is a moderately experienced practical organic chemist of graduate or doctoral standing; experimental descriptions are given in sufficient detail for such an individual to be able to repeat the particular transformations cited, and, it is to be hoped, to be able to apply these general procedures to the case in question. Quantities are normally defined in molar proportions. Most reactions—and, in practice, all reactions—should be performed in a nitrogen or argon atmosphere: carbon–silicon bonds are relatively stable, but nitrogen–, oxygen–, and halogen–silicon bonds are hydrolytically labile to varying degrees; additionally, reactions involving organolithium species should preferably be carried out under argon. All reaction solvents should be appropriately dried—for example, ether and THF from sodium/benzo-phenone ketyl radical. For many reactions, the apparatus described by Seebach (1) is highly recommended.

It has also been assumed that the reader/practitioner has ready access to *Organic Syntheses*, *Organic Reactions* and *Organometallic Syntheses*: detailed descriptions given there are not, as a rule, repeated here. Chapters 18, 19 and 20 list the main organosilicon preparations reported recently in these series.

REFERENCE

1. D. Seebach and K.-H. Geiss, *J. Organometal. Chem. Library* **1**, 1 (1976).

Detailed Contents

19. Organic Reactions

20. Organometallic Syntheses

Abbreviations

Ac	acetyl
AcOH	acetic acid
Ar	aryl
18-Crown-6	1,4,7,10,13,16-hexaoxacyclooctadecane
DABCO	1,4-diazabicyclo[2.2.2]octane
DBN	1,5-diazabicyclo[4.3.0]non-5-ene
DBU	1,5-diazabicyclo[5.4.0]undec-5-ene
DIBAL	di-isobutylaluminium hydride
DME	1,2-dimethoxyethane
DMF	N,N-dimethylformamide
DMSO	dimethylsulphoxide
E (E$^+$)	electrophile
HMDS	hexamethyldisilazane [bis(trimethylsilyl)amine]
HMPA	hexamethylphosphoramide (CAUTION—CANCER SUSPECT AGENT)
Im	imidazol-1-yl
LDA	lithium di-isopropylamide
LDMAN	lithium 1-(dimethylamino)naphthalenide
mcpba	m-chloroperbenzoic acid
MEM	2-methoxyethoxymethyl
NBS	N-bromosuccinimide
Nu (Nu$^-$)	nucleophile
Ph	phenyl
py	pyridine
Red-Al^{TM}	sodium bis(methoxyethoxy)aluminium hydride (^{TM}Aldrich)
TBAF	tetra-n-butylammonium fluoride
TBDMS	t-butyldimethylsilyl
TBDMSCl	t-butyldimethylsilyl chloride
TBDMSOTf	t-butyldimethylsilyl trifluoromethanesulphonate
TBDPS	t-butyldiphenylsilyl
TDSCl	thexyldimethylsilyl chloride
TDSOTf	thexyldimethylsilyl trifluoromethanesulphonate
TES	triethylsilyl
TESH	triethylsilane
TFA	trifluoroacetic acid
TfOH	trifluoromethanesulphonic acid

THF	tetrahydrofuran
THP	tetrahydropyranyl
TIPS	tri-isopropylsilyl
TIPSOTf	tri-isopropylsilyl trifluoromethanesulphonate
TMEDA	N,N,N',N'-tetramethylethylenediamine
TMS	trimethylsilyl
TMSCl	trimethylsilyl chloride (chlorotrimethylsilane)
TMSCN	trimethylsilyl cyanide (cyanotrimethylsilane)
TMSI	trimethylsilyl iodide (iodotrimethylsilane)
TMSOTf	trimethylsilyl trifluoromethanesulphonate
Ts	tosyl (p-toluenesulphonyl)

– 1 –

Introduction

The burgeoning use of silicon in organic synthesis is a consequence of certain unique properties of this element. Perhaps paradoxically, although it belongs to the same periodic group as carbon, and naturally shares its quadricovalency, it is as a substitute for hydrogen that it has found its greatest application and utility. Three main factors are involved in this (formal) replacement. These are silicon's bond strengths to other elements, its relative electronegativity, and its ability to temporarily expand its covalency to enhance reactivity; in terms of mainstream chemistry, it cannot usefully indulge in multiple bonding.

1.1. BOND DISSOCIATION ENERGIES AND BOND LENGTHS

Si—C	$318 \, \text{kJ mol}^{-1}$	$1.89 \, \text{Å}$	C—C	$334 \, \text{kJ mol}^{-1}$	$1.54 \, \text{Å}$
Si—O	531	1.63	C—O	340	1.41
Si—Cl	471	2.05	C—Cl	335	1.78
Si—F	808	1.60	C—F	452	1.39

From even this limited thermochemical data, it can be seen that, relative to carbon, silicon makes *strong* bonds to O, to F and to Cl.

1.2. RELATIVE ELECTRONEGATIVITY

C	N	O	F
2.35	3.1	3.5	4.0

Si	P	S	Cl
1.64	2.1	2.5	2.8

Regardless of the scale used, silicon always appears markedly more electropositive than carbon, resulting in polarization of silicon–carbon bonds in the sense $Si^{\delta+}$—$C^{\delta-}$, and for nucleophilic attack to occur at silicon.

1

1.3. CLEAVAGE OF SILICON–CARBON
AND SILICON–OXYGEN BONDS

The silicon–carbon bond is quite stable towards homolytic fission, but it is readily cleaved by ionic reagents, either by initial nucleophilic attack at Si or by electrophilic attack at C. Since C—H bonds formally dissociate in the same direction, $C^- H^+$, as do C—Si bonds, $C^- Si^+$, a good indication of the likely behaviour of a silicon–carbon bond can be predicted by consideration of an analogous hydrogen–carbon bond. Just as Ar—H bonds are cleaved by electrophiles such as Br_2, so are Ar—Si bonds. Similarly, the β-elimination reactions displayed by H—C—C—X systems occur even more readily in the fragmentation reactions of Si—C—C—X systems, as exemplified by Peterson olefination (Chapter 10). As a broad generalization, it is usually the case that when a C—H bond can be cleaved by a particular ionic reagent, then the corresponding $C—SiMe_3$ bond will be cleaved by the same reagent even more readily (*1*) (in a competitive situation, a C—Si bond is the more reactive towards oxygen and halogen nucleophiles/bases, whereas a C—H bond is the more reactive towards carbon and nitrogen nucleophiles/bases). Similar parallels can also be drawn for O—H and O—Si bonds, although with the *opposite* emphasis, i.e. O—H bonds can be cleaved more readily than can O—Si bonds. As a most useful "rule of thumb", Fleming (*2*) has suggested that Si bonded to C should be considered as a "super proton", whereas when bonded to O it should be treated as an "enfeebled proton". Such C—Si bond cleavages must be seen in perspective. Silicon's bond to carbon, while certainly polarized, is only weakly so when compared with those of more typical organometallic compounds. In general, organosilicon compounds, i.e. those with silicon–carbon bonds, are regiostable, and they can be handled readily, often without the necessity (but normally the preference) for inert atmospheres or moisture exclusion. The silicon–carbon

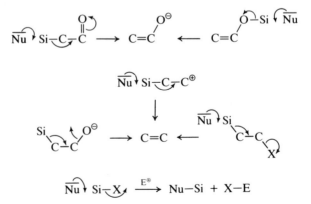

bond can withstand varied reaction conditions, yet it has a latent lability, which can be revealed at an appropriate moment, as will be demonstrated in subsequent chapters. Some characteristic reactions are summarized on p. 2. As mentioned earlier, all of these reactions normally occur *more* readily than do the corresponding hydrogen (or, in the last case, carbon) analogues.

1.4. THE β-EFFECT AND α-ANIONS

Two other important properties of silicon–carbon bonds are that carbonium ions β and carbanions (or metalloid equivalents) α to silicon are favoured over alternatives, i.e. that situations involving Si—C—C$^+$ and Si—C$^-$ are thermodynamically relatively good.

The first of these phenomena is known as the β-effect (3), and can be represented as follows:

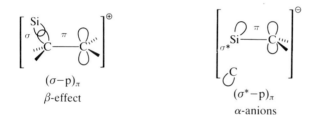

$(\sigma-p)_\pi$

β-effect

$(\sigma^*-p)_\pi$

α-anions

It has been ascribed to $(\sigma-p)\pi$ overlap between the bonding σ-level of the C—Si bond (which will have a relatively high coefficient on carbon owing to its greater electronegativity) with the adjacent empty p-orbital of the carbonium ion; owing to the higher energy of the C—Si bond, there is a better energy match, and therefore greater overall lowering in energy, for this π-overlap than would be obtained from similar C—C or C—H σ-bond hyperconjugation. For significant π-overlap, the C—Si σ-bond must be able to adopt a coplanar relationship with the empty p-orbital; accordingly, operation of the β-effect is seen most clearly in acyclic situations. Additionally, its involvement not only encourages carbonium-ion development but also serves to labilize the C—Si bond involved. The favouring of α-anionoid development can be represented similarly. It has been ascribed to $(\sigma^*-p)\pi$ overlap between the antibonding σ^* level of the C—Si bond (which will have a relatively high coefficient on silicon) with the adjacent filled p-orbital of the carbanion, or highly polarized carbon–metal bond.

One thing that silicon cannot do with any great success is form multiple bonds; compounds with Si=Si or Si=C bonds are quite rare and normally very reactive (*4*). In other words, silicon, in its ground state, greatly prefers to make four single covalent bonds; when undergoing reaction, valence-shell expansion allowing hypercovalency can occur, but still with the involvement of single bonds.

REFERENCES

1. C. Eaborn and R. W. Bott, Synthesis and reactions of the silicon–carbon bond, *Organometallic Compounds of the Group IV Elements,* ed. A. G. MacDiarmid, Vol. 1, Part 1. Marcel Dekker, New York (1968).
2. I. Fleming, Some uses of silicon compounds in synthesis. *Chem. Soc. Rev.* **10,** 83 (1981).
3. For an excellent discussion and list of references, see S. G. Wierschke, J. Chandrasekhar and W. L. Jorgensen, *J. Am. Chem. Soc.* **107,** 1496 (1985).
4. A. G. Brook and K. M. Baines, *Adv. Organometal. Chem.* **25,** 1 (1986); G. Raabe and J. Michl, *Chem. Rev.* **85,** 419 (1985).

$-2-$

Review Articles and Suppliers

2.1. GENERAL REVIEWS (POST-1975)

1. P. F. Hudrlik, Organosilicon compounds in organic synthesis, *J. Organometal. Chem. Library* 1, 127 (1976).
2. E. W. Colvin, Silicon in organic synthesis, *Chem. Soc. Rev.* 7, 15 (1978).
3. I. Fleming, *Comprehensive Organic Chemistry*, Vol. 3, ed. D. H. R. Barton and W. D. Ollis, p. 541. Pergamon, Oxford, 1979.
4. I. Fleming, Some uses of silicon compounds in synthesis, *Chem. Soc. Rev.* 10, 83 (1981).
5. E. W. Colvin, *Silicon in Organic Synthesis*. Butterworths, London, 1981; revised edition, Krieger Press, Florida, 1985.
6. P. D. Magnus, T. Sarkar and S. Djuric, *Comprehensive Organometallic Chemistry*, Vol. 7, ed. G. Wilkinson, F. G. A. Stone and E. W. Abel, p. 515. Pergamon, Oxford, 1982.
7. L. A. Paquette, *Science* 217, 793 (1982).
8. W. P. Weber, *Silicon Reagents for Organic Synthesis*. Springer-Verlag, Berlin, 1983.
9. G. M. Rubottom, *Organometal. Chem. Rev.* 11, 267 (1981); 13, 127 (1982); G. L. Larson, *Organometal. Chem. Rev.* 14, 267 (1984); 17, 1 (1985); *J. Organometal. Chem.* 274, 29 (1984); 313, 141 (1986).
10. E. W. Colvin, Preparation and use of organosilicon compounds in organic synthesis, *Chemistry of the Metal–Carbon Bond*, Vol. 4, ed. S. Patai and F. R. Hartley, Chap. 6, pp. 540–621. Wiley, 1987.

2.2. MORE SPECIFIC, BUT STILL WIDE-RANGING, REVIEWS

11. L. Birkofer and O. Stuhl, Silylated synthons, *Top. Curr. Chem.* 88, 33 (1980).
12. R. Calas, Thirty years in organosilicon chemistry, *J. Organometal. Chem.* 200, 11 (1980).
13. P. Magnus, Organosilicon reagents for carbon–carbon bond-forming reactions, *Aldrichimica Acta* 13, 43 (1980).
14. A. G. Brook and A. R. Bassindale, Molecular rearrangements of organosilicon compounds, *Rearrangements in Ground and Excited States*, Vol. 2, ed. P. de Mayo, pp. 149–227. Academic Press, New York, 1980.
15. D. J. Ager, Silicon-containing carbonyl equivalents, *Chem. Soc. Rev.* 11, 493 (1982).
16. M. Lalonde and T. H. Chan, Use of organosilicon reagents as protective groups in organic synthesis, *Synthesis* 817 (1985).
17. H. Sakurai (ed.), *Organosilicon and Bioorganosilicon Chemistry—Structure, Bonding, Reactivity, and Synthetic Application*. Ellis Horwood, Chichester, 1985.

5

2.3 LISTINGS OF ORGANOSILICON COMPOUNDS

18. V. Bazant, V. Chvalovský and J. Rathouský, *Organosilicon Compounds,* Vols 1 and 2, Academic Press, New York and London, 1965; V. Bazant *et al., Handbook of Organosilicon Compounds,* Vols 1–4, Marcel Dekker, New York, 1973; V. Chvalovský and J. Rathouský, *Organosilicon Compounds,* Vols 5 and 6, Institute of Chemical Process Fundamentals of the Czechoslovak Academy of Sciences, Prague, 1977.

19. D. R. M. Walton, Organosilicon compounds, *Dictionary of Organometallic Compounds,* Vol. 2. Chapman and Hall, London, 1984.

2.4. SUPPLIERS OF ORGANOSILICON COMPOUNDS

Aldrich Chemical Co., Milwaukee, Wisconsin, USA, and Gillingham, Dorset SP8 4JL, UK, and subsidiary companies.

Fluka AG, CH-9470 Buchs, Switzerland.

K and K Greef Chemicals, Croydon CR9 3QL, UK.

P.C.R. Research Chemicals, Inc., Gainsville, Florida, USA.

Petrarch Systems, Inc., Levittown, Pennsylvania, USA.

Pierce Chemical Co., Rockford, Illinois, USA, and Pierce and Warriner (UK) Ltd, Chester, Cheshire, UK.

Ventron GmbH, Karlsruhe, West Germany.

-3-

Vinylsilanes

The considerable synthetic utility of vinylsilanes (*1*) is governed by the availability of suitable stereoselective routes. Most existing methodologies start from either alkynes, carbonyl compounds or vinyl halides.

3.1. PREPARATION

3.1.1. From Alkynes

Alkynes provide one of the most fruitful sources of vinylsilanes. Terminal and internal alkynes can be treated directly with silyl cuprates, or with hydridosilanes (hydrosilylation—see Chapter 17); problems of regioselectivity arise in internal cases. Alternatively, terminal alkynes can be converted into the corresponding alkynylsilanes (Chapter 7); such species react regioselectively with a wide range of organometallic reagents, and they can also be reduced catalytically. These processes are summarized here for terminal alkynes:

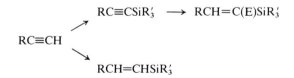

3.1.1.1. Silametallation of terminal alkynes

Fleming has shown (*2*) that the cuprate reagent (Chapter 8) derived from dimethylphenylsilyl lithium and copper(I) cyanide (molar ratio 2:1) adds regioselectively in an overall *syn* manner to terminal alkynes, the silyl moiety becoming attached to the terminal carbon atom (variation in reagent

7

molar ratios can produce isomeric products):

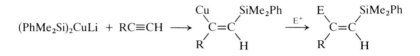

The intermediate vinyl copper species react readily with a variety of electrophiles, with retention of stereochemistry.

(E)-1-Dimethylphenylsilylhex-1-ene (2)

$(PhMe_2Si)_2CuLi$ + $n-C_4H_9C{\equiv}CH$ $\longrightarrow$ (E)-n-$C_4H_9CH{=}CHSiMe_2Ph$

A solution of hex-1-yne (4.5 mmol) in THF (1 ml) was added slowly to lithium bis(phenyldimethylsilyl)cuprate (Chapter 8) (5 mmol, based on CuCN) at 0 °C, and the mixture was stirred for 15 min at 0 °C. Saturated ammonium chloride solution (1 ml) was added, and stirring was continued for 5 min at 0 °C. Light petroleum was then added, and the organic layer was washed with ammonium chloride solution, and dried. Concentration and chromatographic purification on silica gel gave the vinylsilane (4.23 mmol, 94%).

There are, however, two disadvantages associated with use of the phenyldimethylsilyl group. Based on the reaction stoichiometry, for each equivalent of substrate, one silyl group is unused, and after work-up this appears as a relatively involatile by-product. Secondly, after synthetic use of such vinylsilanes involving desilylation, a similar problem of by-product formation arises. One solution to these problems lies in the use of the tri-methylsilyl group (Chapter 8), since the by-product, hexamethyldisiloxane, is volatile and normally disappears on work-up.

Both silylmagnesium and silylaluminium species add to terminal alkynes in the presence of a transition-metal catalyst with high regio- and stereo-selectivity. For example, platinum-catalysed silamagnesiation (*3*) followed by aqueous quenching provides exclusively (*E*)-1-silylalk-1-enes.

(E)-1-Dimethylphenylsilyldodec-1-ene (3)

$n-C_{10}H_{21}C{\equiv}CH$ $\longrightarrow$ (E)-n-$C_{10}H_{21}CH{=}CHSiMe_2Ph$

An ethereal solution of MeMgI (1.1 M, 2 mmol) was added to a solution of dimethylphenylsilyl lithium (0.52 M, 2 mmol) in THF at 0 °C. After stirring for 15 min, *cis*-PtCl$_2$(n-Bu$_3$P)$_2$ (0.01 mmol) was added, and the resulting solution was stirred for an additional 15 min. A solution of dodec-1-yne (1 mmol) in THF (3 ml) was added, and stirring was continued for 30 min at room temperature. Work-up (ether, 1 N HCl) and purification by silica-gel chromatography gave the vinylsilane (0.9 mmol).

3.1.1.2. Hydrosilylation

The addition of hydridosilanes across the triple bond of alkynes can be accomplished using a variety of catalysts, the best being chloroplatinic acid. *cis*-Addition, with terminal regioselectivity, is normally observed, with lower temperatures favouring a greater proportion of the terminal isomer (see also Chapter 17).

(E)-1-Trimethylsilyloct-1-ene (4)

$$n\text{-}C_6H_{13}C\equiv CH \quad \xrightarrow[\text{H}_2\text{PtCl}_6]{\text{Et}_3\text{SiH}} \quad (E)\text{-}n\text{-}C_6H_{13}CH\!=\!CHSiEt_3$$

To a stirred ice-cold mixture of oct-1-yne (18 mmol) and Et$_3$SiH (26 mmol) were added six drops of a solution of H$_2$PtCl$_6$ in propan-2-ol (0.1 M). Stirring was continued for 4 h at 0 °C, then for 4 h at ambient temperature. Direct distillation (Kugelrohr) gave a mixture of vinylsilanes (14 mmol, 77%), b.p. 165–175 °C/25 mmHg: ^{1}H n.m.r. integration indicated a terminal : internal regioisomeric ratio of 4 : 1.

A higher degree of regioselectivity can be attained by the use of lower temperatures, as exemplified by the hydrosilylation of hex-1-yne with trichlorosilane (5).

(E)-1-Trichlorosilylhex-1-ene (5)

$$n\text{-}C_4H_9C\equiv CH \quad \xrightarrow[\text{H}_2\text{PtCl}_6]{\text{Cl}_3\text{SiH}} \quad (E)\text{-}n\text{-}C_4H_9CH\!=\!CHSiCl_3$$

A mixture of hex-1-yne (80 mmol), trichlorosilane (8 ml) and a solution of H$_2$PtCl$_6$ in propan-2-ol (5 drops, 0.1 M) was kept at 2 °C for 15 h, by which time hydrosilylation was complete. Direct distillation gave a mixture of vinylsilanes (60 mmol, 75%). G.l.c. indicated a 95 : 5 mixture of terminal : internal regioisomers.

Notes. This method can be made fully reproducible by using a pre-activated catalyst. For a 100 mmol scale reaction, the above catalyst solution (several drops) was mixed with the alk-1-yne (several mmol) and trichlorosilane (several mmol), and the total heated under reflux for *ca.* 1 h. After cooling, the remaining alkyne and trichlorosilane were added, and the solution kept at ambient temperature or at 2 °C for 15–20 h.

Regiospecificity of addition becomes reversed, for obvious steric reasons, when alkynylsilanes (Chapter 7) are employed as substrates: the product 1,2-disilylalkynes undergo regiospecific protiodesilylation, and so provide a good route to 2-trimethylsilylalk-1-enes, as is discussed later.

3.1.2. From Alkynylsilanes

Hydrometallation and carbometallation of alkynylsilanes proceeds regio- and stereospecifically, the metal becoming attached to the silicon-bearing carbon atom in what is normally a *cis*-addition process (hydrostannylation, however, shows the opposite regioselectivity). Electrophilic cleavage, with retention, of the carbon–metal bond then leads to vinylsilanes of various types.

3.1.2.1. Hydrostannylation

Stannylation, followed by trans-metallation and subsequent electrophilic quenching, provides a good route (*6*) to simple functionalized vinylsilanes:

$$Me_3SiC{\equiv}CH \longrightarrow Me_3SiCH{=}CHE$$

(E)-1-Trimethylsilyl-2-tri-n-butylstannylethene (6)

$$Me_3SiC{\equiv}CH \ + \ n\text{-}Bu_3SnH \longrightarrow (E)\text{-}n\text{-}Bu_3SnCH{=}CHSiMe_3$$

A mixture of Bu₃SnH (0.2 mol) and trimethylsilylethyne (Chapter 7) (0.25 mol) was heated at 100 °C for 83 h. Distillation afforded the stannylethene (0.196 mol), b.p. 96–102 °C/0.5 mmHg.

Transmetallation. A solution of n-BuLi (11 mmol, 1.5 M in hexane) was added to a pre-cooled (−70 to −75 °C) solution of the stannylethene (10 mmol) in THF (30 ml) with stirring. After 1 h at this temperature, the solution was allowed to warm to −30 °C over 0.5 h, then recooled to −78 °C.

(E)-3-Trimethylsilylpropenoic acid (6)

$$(E)\text{-}Me_3SiCH{=}CHLi \ + \ CO_2 \longrightarrow (E)\text{-}Me_3SiCH{=}CHCOOH$$

To a cooled ($-110\,^\circ$C) solution of the silylvinyl lithium (from the stannylethene (37 mmol) and n-BuLi (41 mmol)) in THF (100 ml) was added precooled crushed CO_2 (*ca.* 150 ml). The mixture was stirred at $-110\,^\circ$C for 0.5 h, and then allowed to warm to ambient temperature overnight. The mixture was poured into ether (200 ml) and extracted thoroughly with aqueous NaOH (1 M). The basic extracts were acidified with dilute HCl at $0\,^\circ$C and extracted well with ether. The ethereal extracts were dried, concentrated, and the residue was distilled to give the title acid (32 mmol, 86%), b.p. 68–74 $^\circ$C/0.5 mmHg.

3.1.2.2. Hydroalumination

This occurs in a *trans*-manner in hydrocarbon solvents, while in donor solvent mixtures, clean *cis*-addition takes place. The resulting vinylalanes undergo ready carbodemetallation as the corresponding alanates.

(Z)-3-Trimethylsilyl-3-decene (7)

$$n\text{-}C_6H_{13}C\equiv CSiMe_3 \longrightarrow \underset{H}{\overset{n\text{-}C_6H_{13}}{}}C=C\underset{C_2H_5}{\overset{SiMe_3}{}}$$

In a solution of 1-trimethylsilyloctyne (10 mmol) in ether (20 ml) was added DIBAL (10 mmol) with stirring at $-20\,^\circ$C. The reaction mixture was heated under reflux for 2 h, cooled to $0\,^\circ$C, and treated with MeLi (11 mmol, 1.1 M in ether). It was brought to ambient temperature over 0.5 h, and stirred for a further 1 h, after which it was cooled to $-78\,^\circ$C, and the complex CuI.(EtO)$_3$P (10 mmol) in THF (50 ml) added. To the resulting dark brown mixture was added EtI (16.5 mmol) at $-78\,^\circ$C. The reaction mixture was allowed to warm to ambient temperature over 18 h. It was then acidified with dilute HCl/ice, extracted with ether, washed with saturated ammonium chloride solution brine, and dried. Concentration followed by chromatography on silica gel, eluting with hexane, gave the vinylsilane (7.8 mmol, 78%).

Simple protonolysis of the vinylalane intermediates produces (Z)-2-alkynylvinylsilanes: these species can be readily and cleanly isomerized to the corresponding (*E*)-isomers.

(Z)-1-Cyclohexyl-2-trimethylsilylethene (8)

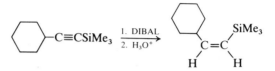

The hexane solvent was removed from a solution of DIBAL (22 mmol, 1 M in hexane) at reduced pressure and at ambient temperature, and ether (10 ml) was introduced. 1-Cyclohexyl-2-trimethylsilylethyne (20 mmol) was added at such a rate as to maintain ambient temperature within the reaction, and, after 15 min, the reaction flask was placed in a preheated (40 °C) bath for 1 h. The resulting clear solution was transferred by means of a double-ended syringe to a vigorously stirred cold solution of HCl (50 ml, 10%). The flask was rinsed with ether (20 ml), and the mixture was stirred until the resulting phases were almost clear. The layers were separated, and the aqueous layer was extracted with ether (40 ml). The combined organic extracts were washed successively with dilute HCl (20%), saturated sodium hydrogen carbonate solution and brine, and dried.

Isomerization to the (E)-isomer (8)

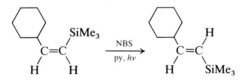

To the above ethereal solution (*ca.* 90 ml) was added pyridine (20 mmol). Then, under irradiation with a UV sunlamp (275 W), the mixture was treated at ambient temperature (water-bath cooling as required) with portions of NBS (5 mol%) at 15 min intervals over a 45 min period. The reaction mixture was decanted from a gummy residue, washed with dilute HCl (50 ml, 3 M), aqueous $CdCl_2$ (20%, to remove traces of pyridine), dilute NaOH (1 M), brine, and dried. Concentration and distillation gave (*E*)-1-cyclohexyl-2-trimethylsilylethene   (17.6 mmol,   88%),   b.p. 66–67 °C/ 4 mmHg.

3.1.3. From Carbonyl Compounds—Sulphonylhydrazone Route

When treated with four equivalents of n-BuLi, arylsulphonylhydrazones give rise to vinyl carbanions/carbanionoids. These species can be trapped (*9*) with trimethylsilyl chloride to produce vinylsilanes in which the silyl group is bonded to what was originally the carbonyl carbon atom:

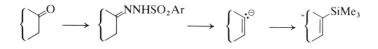

(E)-4-Trimethylsilylhept-3-ene (9)

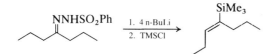

To a stirred solution of the benzenesulphonylhydrazone of heptan-4-one (20 mmol) in TMEDA (90 ml), cooled to −78 °C, was added n-BuLi (80 mmol, 2.4 M in hexane) dropwise. After 5 min, the reaction mixture was allowed to reach ambient temperature, and left for 40 min. TMSCl (98 mmol), was added dropwise to this mixture and stirring was continued for 1.5 h. The resulting mixture was then diluted with dichloromethane, washed successively with dilute HCl, sodium hydrogen carbonate solution (10%) and brine, and dried. Concentration and distillation gave (*E*)-4-trimethylsilylhept-3-ene (12.6 mmol, 60%), b.p. 86–87 °C/47 mmHg.

3.1.4. From Vinyl Halides

One of the most direct routes to vinylsilanes uses vinyl halides as starting materials. Metal–halogen exchange, followed by electrophilic attack by TMSCl, can often provide the vinylsilane quickly and in good yield. As an added bonus, vinyl bromides have been shown (*10, 11*) to proceed through this sequence with retention of double-bond stereochemistry.

(E)-1-Trimethylsilylhex-1-ene (10)

To a stirred solution of (*E*)-1-bromo-1-hexene (10 mmol) in THF/Et₂O/ pentane (24 ml, 4 : 1 : 1), cooled to −120 °C, was added t-BuLi (20 mmol in pentane) over 20 min. The pale yellow solution was stirred for 2 h at −110 to

−120 °C, and then allowed to warm to −78 °C. TMSCl (12 mmol) was added, and stirring was continued for 5 min at −78 °C, and then for 45 min at ambient temperature. The reaction mixture was poured on to ether and brine, and the aqueous layer was extracted with ether. The combined organic extracts were washed with brine and dried. Concentration and distillation gave (*E*)-1-trimethylsilylhex-1-ene (4.6 mmol, 46%), b.p. 80–90 °C/21 mmHg.

(Z)-4-Trimethylsilyloct-4-ene (11)

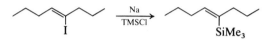

To a stirred suspension of sodium wire (105 mmol, freshly drawn) in ether (100 ml) was added TMSCl (63 mmol). The resulting mixture was stirred at ambient temperature for 15 min, and then a solution of (*Z*)-4-iodo-oct-4-ene (42 mmol, prepared by the reaction of oct-4-yne with HI, isomeric purity 95% (g.l.c.)) in ether (50 ml) was added dropwise. The resulting mixture, which turned blue, was stirred at ambient temperature for 24 h, and then quickly filtered through a pad of Celite, and the pad washed with ether. The filtrate was washed with saturated sodium hydrogen carbonate solution, water, and dried. Concentration and distillation afforded (*Z*)-4-tri-methylsilyloct-4-ene (28 mmol, 67%), b.p. 73–75 °C/18 mmHg, isomeric purity 91.6% (g.l.c.).

3.2. REACTIONS

Vinylsilanes react readily with a range of electrophiles to give products of substitution (*1*). The overall stereochemistry of such substitution will depend on a number of factors, including the stereochemistry of addition and subsequent elimination when 1,2-adducts are discrete species. However, the *regiochemistry* of substitution is normally unambiguous, the β-effect ensuring that carbonium-ion development on attack by the electrophile will occur at the carbon terminus remote, i.e. β, to silicon:

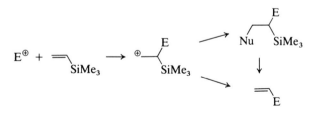

The few exceptions to this general rule arise when the α-carbon carries a substituent that can stabilize carbonium-ion development well, such as oxygen or sulphur. For example, 1-trimethylsilyl trimethylsilyl enol ethers give products (*12*) derived from electrophilic attack at the β-carbon, and the vinylsilane (**1**) reacts with αβ-unsaturated acid chlorides in a Nazarov cyclization (*13*) to give cyclopentenones such as (**2**); the isomeric vinylsilane (**3**), in which the directing effects are additive, gives the cyclopentenone (**4**):

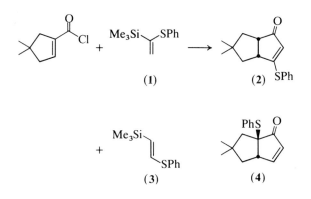

Another exception can be seen in the intriguing case of transannular bromodesilylation (*14*):

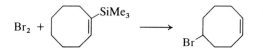

Finally, in a study of Lewis-acid-catalysed intramolecular attack of acetals on vinylsilanes, to produce allylically unsaturated oxacyclics, it has been found (*15*) that the alkene stereochemistry can control the mode of cyclization in an exo- or endocyclic sense, as shown here:

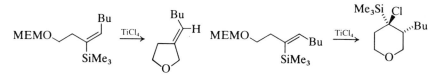

However, such exceptions are relatively rare. Some typical examples of electrophile-induced desilylation are given on p. 16.

3.2.1. Protiodesilylation

(Z)-Tetradec-7-ene (16)

Procedure (i) (16). To a solution of *(E)*-7-trimethylsilyltetradec-7-ene (0.5 mmol) in benzene (1 ml) was added HI (0.04 ml) and the mixture was stirred at ambient temperature for 15 min. After neutralization, chromatography of the reaction mixture gave *(Z)*-tetradec-7-ene (0.45 mmol, 90%). G.l.c. indicated contamination of *ca.* 6% of the *(E)*-isomer.

Procedure (ii) (17). A solution of the vinylsilane (5 mmol) and *p*-toluenesulphinic acid (1 mmol) (*a*) in moist MeCN (15 ml) (*b*) was heated under reflux for 1.5 h. Evaporation followed by filtration in hexane through Al$_2$O$_3$ (20 g) gave the desilylated product (85–92%).

Notes. (*a*) Commercial *p*-toluenesulphinic acid sodium salt was dissolved in water and acidified with dilute sulphuric acid. The precipitated sulphinic acid was filtered off and dried at 20 °C/0.1 mmHg.
(*b*) Water (2%) was added to reagent-grade MeCN.

2-Trimethylsilyloct-1-ene (18)

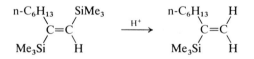

A solution of *(E)*-1,2-bis(trimethylsilyl)oct-1-ene (12 mmol) in glacial AcOH (31 ml) and H$_2$O (1.6 ml) was heated with stirring at 110 °C for 29 h. The mixture was then cooled to 0 °C, aqueous NaOH (60 ml, 9 M) added, and the total was extracted with ether (50 ml). The ethereal extract was washed with saturated sodium hydrogen carbonate solution and brine, and dried. Concentration and distillation gave 2-trimethylsilyloct-1-ene (11.5 mmol, 96%), b.p. 106 °C/1 mmHg.

3.2.2. Deuteriodesilylation

(Z)-7-Deuteriotetradec-7-ene (16)

A mixture of (E)-7-trimethylsilyltetradec-7-ene (1 mmol), benzene (2 ml), D_2O (0.1 ml, 99% pure) and I_2 (0.1 mmol) was heated under reflux for 2 h. Direct chromatographic purification gave (Z)-7-deuteriotetradec-7-ene (0.9 mmol, d_1 89%).

3.2.3. Acylation

1-Acetyl-4,4-dimethylcyclohex-1-ene (19)

A solution of 4,4-dimethyl-1-trimethylsilylcyclohex-1-ene (2 mmol) in dichloromethane (100 ml) was added over 6 h with stirring to a mixture of acetyl chloride (6 mmol) and $AlCl_3$ (6 mmol) in dichloromethane (20 ml). After a further 15 min, the mixture was washed with saturated sodium hydrogen carbonate solution and brine, dried and concentrated *in vacuo*. Preparative t.l.c. on SiO_2, eluting with dichloromethane, gave 1-acetyl-4,4-dimethylcyclohex-1-ene (0.154 mmol, 77%).

3.2.4. Bromination–Desilicobromination, with Inversion of Stereochemistry (20)

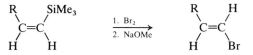

General procedure

To a stirred solution of the vinylsilane (5 mmol) in dichloromethane (5 ml), cooled to $-78\,°C$, was added slowly a solution of Br_2 (6.25 mmol) in dichloromethane (3 ml). To the resulting orange/red solution were added MeOH (125 ml) and Na_2SO_3 (2.5 g), and the resulting mixture was stirred until it became light yellow. While still at $-78\,°C$, the reaction mixture was poured into 10% Na_2SO_3 solution, and shaken until all colour had disappeared. After separation, the aqueous layer was extracted thoroughly with pentane, and the combined organic extracts dried. Excluding light, the solution was concentrated *in vacuo* at ambient temperature to give the crude 1,2-dibromide, which was used immediately. It was quickly dissolved in icecold MeOH (50 ml), and treated with freshly prepared NaOMe in MeOH (7.5 mmol, 1 M) at $0\,°C$. The resulting mixture was stirred at $0\,°C$ for 1 h, and then at ambient temperature for 2 h. It was then partitioned between pentane and H_2O, and the aqueous layer extracted thoroughly with pentane. The combined organic extracts were dried and concentrated at atmospheric pressure in the presence of solid Na_2CO_3. The residue was distilled *in vacuo,* again in the presence of solid Na_2CO_3, to give the vinyl bromide as a colourless liquid (75–93%, *ca.* 97% stereoselectivity).

REFERENCES

1. *Reviews*—T. H. Chan and I. Fleming, *Synthesis* 761 (1979); D. J. Ager, *Chem. Soc. Rev.* **11**, 493 (1982); Z. Parnes and G. I. Bolestova, *Synthesis* 991 (1984).
2. I. Fleming and T. W. Newton, *J. Chem. Soc. Perkin Trans. I* 1805 (1984).
3. H. Hayama, M. Sato, S. Kanemoto, Y. Morizawa, K. Oshima and H. Nozaki, *J. Am. Chem. Soc.* **105**, 4491 (1983).
4. E. W. Colvin, unpublished.
5. K. Tamao, J.-I. Yoshida, H. Yamamoto, T. Kakui, H. Matsumoto, M. Takahashi, A. Kurita, M. Murata and M. Kumada, *Organometallics* **1**, 355 (1982).
6. R. F. Cunico and F. J. Clayton, *J. Org. Chem.* **41**, 1480 (1976).
7. F. E. Ziegler and K. Mikami, *Tetrahedron Lett.* **25**, 131 (1984).
8. G. Zweifel and H. P. On, *Synthesis* 803 (1980).
9. T. H. Chan, A. Baldassarre and D. Massuda, *Synthesis* 801 (1976). See also R. T. Taylor, C. R. Degenhardt, W. P. Melega and L. A. Paquette, *Tetrahedron Lett.* 159 (1977); A. R. Chamberlin, J. F. Stemke and F. T. Bond, *J. Org. Chem.* **43**, 147 (1978).
10. H. Neumann and D. Seebach, *Tetrahedron Lett.* 4839 (1976); *Chem. Ber.* **111,** 2785 (1978).
11. P. F. Hudrlik, A. K. Kulkarni, S. Jain and A. M. Hudrlik, *Tetrahedron* **39**, 877 (1983).
12. I. Kuwajima, M. Kato and T. Sato, *J. Chem. Soc. Chem. Commun.* 478 (1978); T. Sato, T. Abe and I. Kuwajima, *Tetrahedron Lett.* 259, 1383 (1978); N. Minami, T. Abe and I. Kuwajima, *J. Organometal. Chem.* **145,** C1 (1978).
13. P. Magnus, D. A. Quagliato and J. C. Huffman, *Organometallics* **1**, 1240 (1982); P. Magnus and D. Quagliato, *J. Org. Chem.* **50**, 1621 (1985). For related Nazarov studies, see T. K. Jones and S. E. Denmark, *Helv. Chim. Acta* **66**, 2377 (1983);

G. Kjeldsen, J. S. Knudsen, L. S. Ravn-Petersen and K. B. G. Torsell, *Tetrahedron* **39**, 2237 (1983).
14. D. Dhanak, C. B. Reese and D. E. Williams, *J. Chem. Soc. Chem. Commun.* 988 (1984).
15. L. E. Overman, A. Castañeda and T. A. Blumenkopf, *J. Am. Chem. Soc.* **108**, 1303 (1986).
16. K. Utimoto, M. Kitai and H. Nozaki, *Tetrahedron Lett.* 2825 (1975).
17. G. Büchi and H. Wüest, *Tetrahedron Lett.* 4305 (1977).
18. P. F. Hudrlik, R. H. Schwartz and J. C. Hogan, *J. Org. Chem.* **44**, 155 (1979).
19. I. Fleming and A. Pearce, *J. Chem. Soc. Perkin Trans. I* 2485 (1980).
20. R. B. Miller and G. McGarvey, *J. Org. Chem.* **44**, 4623 (1979).

$-4-$

$\alpha\beta$-Epoxysilanes

Vinylsilanes (Chapter 3) can be readily converted into $\alpha\beta$-epoxysilanes, normally by treatment with mcpba (*1*). Alternatively, α-chloro-α-lithio-α-trimethylsilanes react efficiently with aldehydes and ketones in a manner reminiscent of the Darzens reaction (*2*).

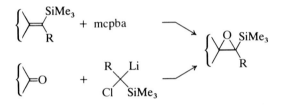

4.1. PREPARATION

General procedure using mcpba (3)

A solution of mcpba (11.5 mmol) in dichloromethane (30 ml) was added to a stirred solution of the vinylsilane (10 mmol) in dichloromethane (50 ml) at 0 °C. After stirring for 1 h, the mixture was washed with aqueous sodium hydrogen sulphite (50 ml), saturated sodium hydrogen carbonate solution and brine. The organic solution was then dried and concentrated, prior to purification by distillation or chromatography.

4.2. EPOXIDATION AND HOMOLOGATION OF CARBONYL COMPOUNDS

$\alpha\beta$-Epoxysilanes are of considerable synthetic utility, in that they undergo an acid-catalysed rearrangement (*4*) to yield carbonyl compounds, with the carbonyl carbon being the one that originally carried the silyl group:

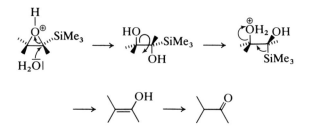

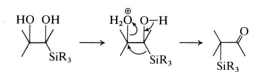

The above mechanism involves α-opening of the epoxysilane, followed by a 1,2-elimination of a β-hydroxysilane (Peterson olefination, Chapter 10). However, it has recently been shown that αβ-dihydroxysilanes, particularly t-butyldimethylsilyl species, undergo an acid-catalysed sila-pinacol rearrangement to produce β-aldehydo- and β-ketosilanes (5):

These silanes, especially when trimethylsilyl, as is normally the case, undergo a facile solvolytic loss of the silyl group to give the parent carbonyl compound.

Cyclohexane carboxaldehyde (2)

A solution of s-BuLi (1.05 mmol, 1.1 M in cyclohexane) was added dropwise to a stirred solution of chloromethyltrimethylsilane (1 mmol) in THF (to give a 1 M solution of reagent) at −78 °C. After 5 min, TMEDA (1.05 mmol) was added, and the stirred solution was allowed to warm to −55 °C over 0.5 h, by which time it was pale yellow (the presence of TMEDA increases the rate of formation of the reagent, but has no other advantages).

Notes. (a) Similar treatment of (1-chloroethyl)trimethylsilane provided the equivalent homologous reagent.

(b) Both reagents were used immediately.

Cyclohexanone (5.7 mmol) was added to a solution of the reagent
(6.2 mmol) at -78 to $-50\,°C$. After 0.5 h at this temperature, it was allowed
to warm to ambient temperature over 3 h. The mixture was poured into
dilute HCl (25 ml, 0.5 M), extracted with dichloromethane (3×30 ml),
and the organic extracts dried and concentrated, to give the epoxysilane
(4.7 mmol, 83%).

To a stirred solution of the epoxysilane (1 mmol) in THF/water (3 ml, 4 : 1)
was added perchloric acid (0.1 ml, 70%), and stirring was continued for 4 h
at ambient temperature. The mixture was then poured into water (20 ml),
and extracted with dichloromethane (3×20 ml). Drying and concentration
gave cyclohexane carboxaldehyde (0.71 mmol, 71%).

4.3. REARRANGEMENT TO SILYL ENOL ETHERS

Geometrically defined $\alpha\beta$-epoxysilanes have been shown (6) to undergo
a highly stereoselective rearrangement to silyl enol ethers (see also
Chapter 15). This rearrangement is catalysed by boron trifluoride etherate,
and seems to involve β-opening of the epoxysilane, as shown:

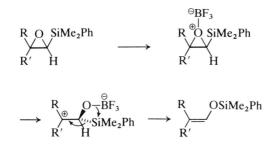

General procedure (6)

Boron trifluoride etherate (1 mmol) was added dropwise to a stirred solution
of the epoxysilane (1 mmol) in dichloromethane (5 ml) at $-78\,°C$, and the
mixture was stirred for 5 min. The reaction mixture was quenched with
saturated sodium hydrogen carbonate solution (1 ml), and allowed to warm
gradually to ambient temperature. The organic phase was washed with brine
(3×5 ml), dried and concentrated. The (Z)-epoxysilane gave the (Z)-silyl
enol ether (68%, 96 : 4 (Z) : (E)), and the (E)-isomer gave the (E)-silyl enol
ether (69%, 95 : 5 (E) : (Z)).

REFERENCES

1. J. J. Eisch and J. T. Trainor, *J. Org. Chem.* **28**, 487 (1963).
2. C. Burford, F. Cooke, G. Roy and P. Magnus, *Tetrahedron* **39**, 867 (1983); F. Cooke, G. Roy and P. Magnus, *Organometallics* **1**, 893 (1982).
3. E.W. Colvin, unpublished.
4. G. Stork and E. Colvin, *J. Am. Chem. Soc.* **93**, 2080 (1971).
5. R. F. Cunico, *Tetrahedron Lett.* **27**, 4269 (1986).
6. I. Fleming and T. W. Newton, *J. Chem. Soc. Perkin Trans. I* 119 (1984).

-5-

Allylsilanes

5.1 PREPARATION

A variety of routes are available for the preparation of allylsilanes (*1*) with the simplest and most direct being the silylation of allyl-metal species. Other routes exemplified in this chapter include Wittig methodology, the use of silyl anions/anionoids in allylic substitution, and hydrometallation of propargylsilanes.

5.1.1. Silylation of Allyl-Metal Species

5-Trimethylsilylcyclopentadiene (2)

Freshly distilled cyclopentadiene (0.5 mol) was added dropwise to a stirred mixture of Na sand (0.5 mol) in THF (150 ml), and stirring was continued for a further 3 h at room temperature. After this time, TMSCl (0.5 mol) was added dropwise over 1 h, and stirring was continued for 3 h. The reaction mixture was poured into cold water (150 ml), and thoroughly extracted with ether. The combined ethereal extracts were concentrated and the residue was distilled, to give the title silane (0.365 mol, 73%), b.p. 41–43 °C/ 16 mmHg.

1-Trimethylsilylcyclopent-2-ene (3)

A mixture of TMSCl (0.51 mol), Mg (0.77 mol) and THF (250 ml) was cooled to 5 °C in an ice bath. An addition funnel was charged with a solution of 3-chlorocyclopentene (0.51 mol) in THF (500 ml), and cooled with a dry ice jacket. The resulting cooled solution was added dropwise to the stirred reaction mixture over several hours. The reaction mixture was allowed to warm to room temperature, and then stirred overnight. It was washed with water (5 × 100 ml), and the aqueous washes back-extracted with pentane (50 ml). The combined organic phases were washed with brine (100 ml), and dried. Concentration and distillation at atmospheric pressure gave the title silane (0.48 mol, 94%), b.p. 140–146 °C.

Note. 1-Trimethylsilylcyclohex-2-ene can be prepared similarly from 3-bromocyclohexene, in 54% yield, b.p. 69–72 °C/10 mmHg.

5.1.2. Reductive Silylation

1,4-Bis(trimethylsilyl)cyclohex-2-ene (4)

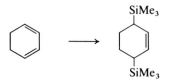

A mixture of powdered Mg (0.1 g atom), cyclohexa-1,3-diene (0.1 mol), and TMSCl (0.35 mol) in HMPA (CAUTION—CANCER SUSPECT AGENT) (100 ml) was heated under reflux with stirring for 65 h. After this time, the Mg had disappeared, and the medium had separated into two phases. On cooling, the mixture was poured on to cold saturated ammonium chloride solution, and thoroughly extracted with ether. The combined ethereal extracts were washed with brine and dried. Concentration and distillation gave the silane (0.065 mol, 65%) as a mixture of stereoisomers, b.p. 121–124 °C/25 mmHg.

5.1.3. Wittig Route to Allylsilanes

General procedure (5)

n-BuLi (25 mmol, 1.66 M in hexane) was added dropwise over 0.5 h to a stirred suspension of methyltriphenylphosphonium bromide (22.5 mmol) in THF (40 ml) at 0 °C. After warming to room temperature, the mixture was stirred for 1 h, recooled to 0 °C, and iodomethyltrimethylsilane (22.5 mmol) was added dropwise over 10 min. After warming to room temperature, the mixture was stirred for 1 h, then cooled to −78 °C, and treated with a second equivalent of n-BuLi (25 mmol, 1.66 M in hexane). After warming to room temperature, the mixture was stirred for a further 1.5 h, to give a dark-red solution of the ylide. The solution was again cooled to −78 °C, and the carbonyl compound (20 mmol) in THF (10 ml) added dropwise over 15 min. After 0.5 h, the mixture was allowed to warm to room temperature, and stirred for a further 16 h. It was then poured into saturated ammonium chloride solution (100 ml), and extracted with ether (3 × 300 ml). The allylsilane was isolated either by chromatography on silica gel, eluting with carbon tetrachloride, or by distillation (60–80%).

5.1.4. S$_N$ Displacement Routes

An extensive study (*6*) of S$_N$ displacement reactions of allyl halides using silyl anions/anionoids has provided the following regioselective alternatives:

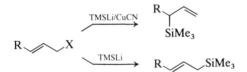

3-Trimethylsilylnon-1-ene (6)

$$\text{n-C}_6\text{H}_{13}\text{CH=CHCH}_2\text{Cl} \xrightarrow{\text{TMSCuLi}} \text{n-C}_6\text{H}_{13}\underset{\underset{\text{SiMe}_3}{|}}{\text{CHCH=CH}_2}$$

To a solution of hexamethyldisilane (2.5 mmol) in HMPA (CAUTION—CANCER SUSPECT AGENT) (3 ml) at 0–5 °C was added methyl lithium (2.5 mmol, 1.5 M MeLi.LiBr complex in ether) dropwise. After being stirred for 3 min, the red solution was treated with CuI (2.5 mmol) in Me$_2$S (1 ml), and the resulting black reaction mixture was stirred for 3 min. Ether (6 ml) was added, the reaction mixture was cooled to −60 °C, and stirred for 5 min. A solution of 1-chloronon-2-ene (1 mmol) in ether (1 ml) was added drop-wise, and the mixture stirred for 1 h at −60 to −50 °C. The cold solution was poured into pentane (50 ml) and saturated ammonium chloride solution (50 ml, buffered to pH 8 by the addition of ammonium hydroxide), and the mixture was stirred vigorously for 1 h. The aqueous phase was re-extracted with pentane, and the combined organic extracts were dried and concen-trated. Purification by chromatography on silica gel, eluting with pentane, gave the allylsilane (0.87 mmol, 87%), contaminated with 2% of its 1-trimethylsilyl regioisomer.

1-Trimethylsilylnon-2-ene (6)

$$\text{n-}C_6H_{13}CH{=}CHCH_2Cl \xrightarrow{\text{TMSLi}} \text{n-}C_6H_{13}CH{=}CHCH_2SiMe_3$$

To a solution of hexamethyldisilane (2.5 mmol) in HMPA (CAUTION— CANCER SUSPECT AGENT) (3 ml) at 0–5 °C was added methyl lithium 2.5 mmol, 1.5 M MeLi.LiBr complex in ether) dropwise. After being stirred for 3 min, the resulting red solution was diluted with ether (6 ml), cooled to −60 °C, and stirred for 5 min, when the solution was yellow. A solution of 1-chloronon-2-ene (1 mmol) in ether (1 ml) was added dropwise, and the mixture was stirred for 1 h at −60 to −50 °C. The cold solution was poured into pentane (50 ml) and saturated ammonium chloride solution (50 ml), the aqueous phase was re-extracted with pentane, and the combined organic extracts were dried and concentrated. Purification by chromatography on silica gel, eluting with pentane, gave the allylsilane (0.78 mmol, 78%).

2-Bromo-3-trimethylsilylpropene (6)

$$BrCH_2C(Br){=}CH_2 \xrightarrow{\text{TMSCuLi}} Me_2SiCH_2C(Br){=}CH_2$$

To a solution of hexamethyldisilane (2.5 mmol) in HMPA (CAUTION— CANCER SUSPECT AGENT) (3 ml) at 0–5 °C was added methyl lithium (2.5 mmol, 1.5 M MeLi.LiBr complex in ether) dropwise. After being stirred for 3 min, the red solution was treated with CuI (2.5 mmol) in Me₂S (1 ml), the resulting black reaction mixture was stirred for 3 min, and 2,3-dibromo-propene (1 mmol) was added rapidly *via* a syringe. The reaction mixture was allowed to warm to room temperature, and was stirred for 1.5 h. It was then poured into pentane (25 ml) and saturated ammonium chloride solution (25 ml, buffered to pH 8 by the addition of ammonium hydroxide), and the mixture was stirred vigorously for 1 h. The aqueous phase was re-extracted with pentane, and the combined organic extracts were dried. Removal of

pentane by distillation at atmospheric pressure, followed by distillation
of the residue, gave the allylsilane (0.9 mmol, 90%), b.p. 82–85 °C/
60 mmHg.

Notes. This compound can also be prepared (*7*) from 2,3-dibromopropene
and the complex (TMSCuCN)Li in 63% yield.

Allyl acetates can similarly be transformed into allylsilanes by treatment
with bis(silyl)cuprates (*8*).

(2-Cyclohexylidene-ethyl)trimethylsilane (8)

1-Vinylcyclohexyl acetate (4 mmol) was added to a stirred solution of
lithium bis(trimethylsilyl)cuprate (Chapter 8) (5 mmol) at −23 °C, and kept
at 0 °C for 1 h. The mixture was then diluted with pentane, and saturated
ammonium chloride solution was added. Filtration through glass wool,
washing with saturated ammonium chloride solution, drying, and concentration under reduced pressure gave the crude product. Column chromatography (pentane–ether, 19 : 1 v/v) gave the silane (2.7 mmol, 68%).

5.1.5. Hydrometallation Routes

Monohydroalumination of terminal propargylsilanes (Chapter 7) with two
equivalents of DIBAL proceeds in a stereoselective (*9*), though not regioselective, manner to produce an approximately 1 : 1 mixture of the alkenyl-
alanes (**1**) and (**2**). Protonolysis of the mixture yields the pure (*Z*)-allylsilane
(**3**).

$$RC{\equiv}CCH_2SiMe_3 \longrightarrow
\left[
\begin{array}{ccc}
\underset{H}{\overset{R}{\diagdown}}C{=}C\underset{AlBu_2^i}{\overset{SiMe_3}{\diagup}}
& + &
\underset{Bu_2^iAl}{\overset{R}{\diagup}}C{=}C\underset{H}{\overset{SiMe_3}{\diagdown}}
\\
(1) & & (2)
\end{array}
\right]$$

$$\xrightarrow{H_3O^{\oplus}} \underset{H}{\overset{R}{\diagdown}}C{=}C\underset{H}{\overset{SiMe_3}{\diagup}}$$

$$(3)$$

(Z)-1-Trimethylsilylhept-2-ene (9)

$$C_4H_9C{\equiv}CCH_2SiMe_3 \xrightarrow[\text{2. H}^+]{\text{1. DIBAL}} (Z)\text{-}C_4H_9CH{=}CHCH_2SiMe_3$$

To a solution of 1-trimethylsilylhept-2-yne (15 mmol) in hexane (15 ml) was added DIBAL (30 mmol, neat) with stirring, maintaining the reaction temperature at 25–30 °C by means of a water bath. The solution was stirred at ambient temperature for 30 min, and then heated at 70 °C for 4 h. On cooling to ambient temperature, the reaction mixture was transferred using a double-ended needle to a vigorously stirred mixture of aqueous HCl (30 ml, 3 M), ice (30 g) and pentane (15 ml). The mixture was stirred for a further 15 min, the layers were separated, and the aqueous layer was extracted with pentane (3 × 20 ml). The combined organic extracts were washed with water (25 ml) and brine (25 ml), and dried. Concentration and distillation gave the product (12.75 mmol, 85%), b.p. 71–72 °C/13 mmHg.

Internal propargylsilanes undergo hydroalumination much less successfully. However, hydroboration proceeds without difficulty:

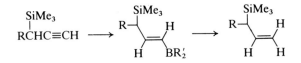

3-Trimethylsilyloct-1-ene (9)

$$\underset{\substack{| \\ \text{n-}C_5H_{11}CHC{\equiv}CH}}{SiMe_3} \longrightarrow \underset{\substack{| \\ \text{n-}C_5H_{11}CHCH=CH_2}}{SiMe_3}$$

To a stirred solution of bis(1,2-dimethylpropyl)borane (5.1 mmol, 1.5 M in THF) was added 3-trimethylsilyloct-1-yne (5 mmol) at 0 °C with stirring. The mixture was stirred at 0 °C for a further 30 min and then at ambient temperature for 30 min. Acetic acid (0.6 ml) was added, and the mixture was heated at 65–70 °C for 2 h. On cooling to 35–40 °C, aqueous sodium acetate (18 mmol, 3 M) and hydrogen peroxide (30%, 1.4 ml, 14 mmol) were added, to complete oxidation of remaining organoboron species. The mixture was stirred at ambient temperature for 30 min, and then saturated with potassium carbonate and extracted with ether (2 × 10 ml). The combined extracts were washed with brine (20 ml), and dried. Concentration and distillation gave the product (3.95 mmol, 79%), b.p. 63–64 °C/4 mmHg.

5.1.6. Carbon-Chiral Allylsilanes by Asymmetric Grignard Cross-Coupling

(R)-1-Trimethylsilyl-1-phenylprop-2-ene (10)

$$\underset{Ph}{\overset{Me_3Si}{>}}\!\!-MgBr \;+\; \overset{}{\diagup}\!\!\diagdown Br \;\longrightarrow\; \diagup\!\!\diagup\overset{SiMe_3}{\underset{H}{\diagdown}}\!\!-Ph$$

To a mixture of vinyl bromide (40 mmol) and the catalyst dichloro-[(R)-N,N-dimethyl-1-[(S)-2-(diphenylphosphino)ferrocenyl]ethylamine]-palladium(II) (0.2 mmol) was added an ethereal solution of [α-(trimethylsilyl)benzyl]magnesium bromide (0.6–1 M, 80 mmol) at −78 °C. The mixture was stirred at 30 °C for 4 days, and then cooled to 0 °C and hydrolysed with dilute aqueous HCl (3 M). The organic layer was separated, and the aqueous layer was re-extracted with ether. The combined organic extracts were washed with saturated sodium hydrogen carbonate solution and water, and dried. Concentration and distillation gave the chiral allylsilane (79%, 66% ee), b.p. 55 °C/0.4 mmHg.

Notes. Lower reaction temperatures give higher enantioselectivities, but with lower chemical yields; in the above case, stirring at 0 °C for four days gave the product in 42% yield but with 95% ee.

5.2 REACTIONS

Allylsilanes, being homologues of vinylsilanes, undergo a similar regio-controlled attack (*1*) by electrophiles, this time at the γ-position, with resulting loss of the silyl group providing products of substitution with a net shift of the double bond:

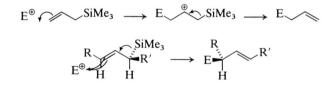

Extensive studies (*11*) of such $S_{E'}$ reactions of allylsilanes have demonstrated a high degree of *anti* stereoselectivity with the majority of electrophiles, except in cases where steric effects play a dominant role.

Allylsilanes are somewhat more reactive than vinylsilanes in such reactions: considering the reaction pathway followed by each substrate on electrophilic attack, i.e. β-carbonium ion development, such a pathway will be expected to have a lower activation energy in the case of allylsilanes. Here, stabilization of the developing positive charge can be more or less continuous, as the β-C—Si bond can overlap with the π-system with no geometrical constraints (at least in acyclic cases). Such orbital overlap will also raise the energy of the HOMO of the π-system, thus increasing its reactivity. Strong evidence that the cation (4) is indeed an intermediate, stabilized by such favourable overlap, has been provided by an investigation (*12*) of the isomeric silanes (5) and (6); protiodesilylation gave the same mixture of products, in the same ratio, in each case. Maximum stabilization with vinylsilanes, on the other hand, requires rotation of the originally coplanar C—Si bond through 90°.

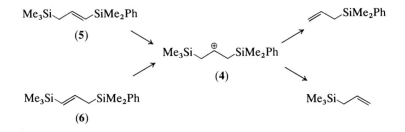

Unlike many other metal-allyl systems, allylsilanes are regio-stable at normal temperatures, with 1,3-sigmatropic shifts occurring at a significant rate only at temperatures in excess of 300 °C. 5-Trimethylsilylcyclopentadiene provides an exception to this generalization, but it can still be handled quite readily.

Certain representative examples of electrophile-induced desilylation of allylsilanes are given below.

5.2.1 Protiodesilylation

General procedure (13)

A solution of the allylsilane (2 mmol) in chloroform (2 ml) was added all at once to boron trifluoride–acetic acid complex (2.2 mmol) with vigorous shaking until a single phase resulted. After 5 min, the solution was poured into saturated sodium hydrogen carbonate solution (10 ml) and extracted with ether (3 × 20 ml). The organic phase was dried and concentrated, and the residue was purified by chromatography or distillation.

5.2.2 Alkylation and Acylation

General procedure (13)

Titanium tetrachloride (5.5 mmol) was added to dichloromethane (5 ml), and the solution was cooled to −78 °C with stirring. This cooled solution was transferred, *via* a syringe, to a stirred solution of the allylsilane (5 mmol) and the alkylating agent (6 mmol) (alkyl chloride, bromide, ethylene oxide, acetyl chloride, or mvk) in dichloromethane, pre-cooled to −78 °C. After 1 h, the reaction mixture was poured into saturated sodium hydrogen carbonate solution (25 ml) and extracted with ether (3 × 50 ml). After drying and concentration, the residue was purified by chromatography or distillation.

Artemisia ketone (14)

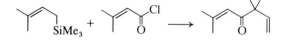

A mixture of senecioyl chloride (0.1 mol) and AlCl₃ (0.1 mol) in dichloromethane (50 ml) was added dropwise over 45 min with stirring to a solution of dimethylallyltrimethylsilane (0.11 mol) in dichloromethane (100 ml), previously cooled to −60 °C. The reaction mixture was maintained at this temperature for 10 min after the final addition, and then poured slowly onto a mixture of crushed ice and ammonium chloride. After washing of the organic layer with brine and drying, artemisia ketone (0.09 mol, 90%), b.p. 87 °C/200 mmHg, was obtained.

Note. Dimethylallylmagnesium chloride was prepared in dilute ether solution at −50 °C. TMSCl was added at ambient temperature, followed by quenching with saturated ammonium chloride at −50 °C. This gave dimethylallyltrimethylsilane (58%), b.p. 100 °C/300 mmHg.

5.2.3. Reaction with Acetals

4-Phenyl-4-methoxybut-1-ene (15)

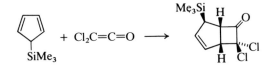

A mixture of TMSOTf (0.1 mmol, 1 mol%), allyltrimethylsilane (11.5 mmol) and dichloromethane (1 ml) was cooled to −78 °C, and to this was added benzaldehyde dimethylacetal (10.5 mmol) in dichloromethane (4 ml). The resulting mixture was stirred for 6 h at −78 °C, and then poured into saturated sodium hydrogen carbonate solution (10 ml) and extracted with ether (3 × 20 ml). The combined organic extracts were washed with brine, dried and concentrated. Chromatography on silica gel (1 : 20 ether : hexane) gave 4-phenyl-4-methoxybut-1-ene (9.2 mmol, 88%).

5.2.4. Cycloaddition of Dichloroketene to 5-Trimethylsilylcyclopentadiene (*16*)

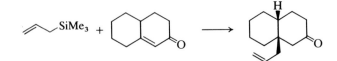

A solution of dichloroacetyl chloride (0.2 mol) in hexane (100 ml) was added dropwise over 2 h to a stirred mixture of 5-trimethylsilylcyclopentadiene (0.15 mol) and triethylamine (0.2 mol) in hexane (300 ml) at 0 °C. After stirring for an additional 3 h at 0–5 °C, the mixture was poured into water (500 ml). The organic layer was separated, dried and concentrated. Distillation gave the allylsilane (0.105 mol, 70%), b.p. 71–73 °C/0.3 mmHg.

5.2.5. Conjugate Addition to Enones

9-Allyl-cis-2-octalone (17)

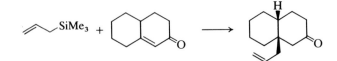

To a solution of $\Delta^{1,9}$-2-octalone (2 mmol) in dichloromethane (3 ml) was added TiCl₄ (2 mmol) dropwise. After 5 min, the solution was cooled to −78 °C, and allyltrimethylsilane (2.8 mmol) in dichloromethane (3 ml) was added dropwise. After stirring for 18 h at −78 °C, and 5 h at −30 °C, water was added to the mixture, which was then extracted thoroughly with ether. The combined organic extracts were washed with water, dried and concentrated. Distillation of the residue gave 9-allyl-*cis*-2-octalone (85%), b.p. 120 °C/5 mmHg.

5.2.6. Reactions of Allylsilane Anions

In general, simple deprotonated allylsilanes react in a γ-regioselective manner with electrophiles, vinylsilanes being obtained as products.

(E)-1-[3-(Trimethylsilyl)allyl]cyclohexan-1-ol (18)

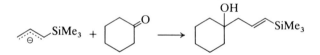

To a stirred solution of s-BuLi (130 mmol, 1.4 M in cyclohexane) and TMEDA (130 mmol) in THF (50 ml), cooled to −78 °C, was added dropwise allyltrimethylsilane (130 mmol). The resulting solution was allowed to warm to −30 °C, and stirred at that temperature for 30 min. Cyclohexanone (51 mmol) was added dropwise (the reaction was complete upon full addition). The reaction mixture was partitioned between dichloromethane (100 ml) and saturated ammonium chloride solution (100 ml), and the separated aqueous layer was extracted with dichloromethane (2 × 50 ml). The combined organic extracts were washed with water (200 ml), dried and concentrated *in vacuo*. Distillation of the residue gave the vinylsilane (36 mmol, 70% based on cyclohexanone), b.p. 66–70 °C/0.025 mmHg.

On the other hand, if the allylsilane anion is first complexed with certain metals, α-regioselectivity then predominates, and a high degree of complementary diastereoselectivity (*19*) can be attained with aldehydes as electrophiles. For example, boron, aluminium and titanium complexation all induce *threo* selectivity whereas the use of tin results in an *erythro*

preference, providing by the Peterson olefination (Chapter 10) stereo-
controlled routes to terminal dienes:

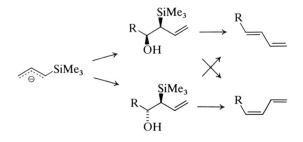

REFERENCES

1. T. H. Chan and I. Fleming, *Synthesis* 761 (1979); Z. Parnes and G. I. Bolestova, *Synthesis* 991 (1984).
2. C. S. Kraihanzel and M. L. Losee, *J. Am. Chem. Soc.* **90,** 4701 (1968).
3. J. M. Reuter, A. Sinha and R. G. Salomon, *J. Org. Chem.* **43,** 2438 (1978) (cf. C. Eaborn, R. A. Jackson and R. Pearce, *J. Chem. Soc. Perkin Trans. I* 2055 (1974)).
4. J. Dunogues, R. Calas, J. Dedier and F. Pisciotti, *J. Organometal. Chem.* **25,** 51 (1970).
5. I. Fleming and I. Paterson, *Synthesis* 446 (1979); D. Seyferth, K. R. Wursthorn and R. E. Mammarella, *J. Org. Chem.* **42,** 3104 (1977).
6. J. G. Smith, S. E. Prozda, S. P. Petraglia, N. R. Quinn, E. M. Rice, B. S. Taylor and M. Viswanathan, *J. Org. Chem.* **49,** 4112 (1984); see also E. Negishi, F. T. Luo and C. L. Rand, *Tetrahedron Lett.* **23,** 27 (1982); B. Laycock, W. Kitching and G. Wickham, *Tetrahedron Lett.* **24,** 5785 (1983); see also Chapter 20 of the present book.
7. B. M. Trost and D. M. T. Chan, *J. Am. Chem. Soc.* **104,** 3733 (1982).
8. I. Fleming and T. W. Newton, *J. Chem. Soc. Perkin Trans. I* 805 (1984). For a three-step sequence that leads to allylsilanes in which the silyl group is at the more substituted end of the double bond, see I. Fleming and D. Waterson, *J. Chem. Soc. Perkin Trans. I* 1809 (1984); see also I. Fleming and D. Marchi, *Synthesis* 560 (1981); I. Fleming and N. K. Terrett, *J. Organometal. Chem.* **264,** 99 (1984); I. Fleming and A. P. Thomas, *J. Chem. Soc. Chem. Commun.* 411 (1985).
9. S. Rajagopalan and G. Zweifel, *Synthesis* 113 (1984).
10. T. Hayashi, M. Konishi, Y. Okamoto, K. Kabeta and M. Kumada, *J. Org. Chem.* **51,** 3772 (1986); for another route to carbon-chiral allylsilanes, see I. Fleming and A. P. Thomas, *J. Chem. Soc. Chem. Commun.* 1456 (1986).
11. T. Hayashi, K. Kabeta, T. Yamamoto, K. Tamao and M. Kumada, *Tetrahedron Lett.* **24,** 5661 (1983) and references therein; I. Fleming and N. K. Terrett, *Tetrahedron Lett.* **24,** 4253 (1983); H. Wetter and P. Scherer, *Helv. Chim. Acta* **66,** 118 (1983); G. Wickham and W. Kitching, *J. Org. Chem.* **48,** 612 (1983). For some intramolecular studies, see S. R. Wilson and M. R. Price, *J. Am. Chem. Soc.* **104,** 1124 (1982); *Tetrahedron Lett.* **24,** 569 (1983).
12. I. Fleming and J. A. Langley, *J. Chem. Soc. Perkin Trans. I* 1421 (1981); see also I. Fleming, D. Marchi and S. K. Patel, *J. Chem. Soc. Perkin Trans. I* 2519 (1981).
13. I. Fleming and I. Paterson, *Synthesis* 446 (1979).

14. J.-P. Pillot, J. Dunoguès and R. Calas, *Tetrahedron Lett.* 1871 (1976).
15. T. Tsunoda, M. Suzuki and R. Noyori, *Tetrahedron Lett.* **21,** 71 (1980).
16. I. Fleming and B.-W. Au-Yeung, *Tetrahedron* **37,** Suppl. 1, 13 (1980).
17. A. Hosomi and H. Sakurai, *J. Am. Chem. Soc.* **99,** 1673 (1977); see also Chapter 18 of the present book.
18. E. Ehlinger and P. Magnus, *J. Am. Chem. Soc.* **102,** 5004 (1980).
19. See for example D. J. S. Tsai and D. S. Matteson, *Tetrahedron Lett.* **22,** 2751 (1981); Y. Yamamoto, Y. Saito and K. Maruyama, *Tetrahedron Lett.* **23,** 4597 (1982); F. Sato, Y. Suzuki and M. Sato, *Tetrahedron Lett.* **23,** 4589 (1982); Y. Yamamoto, Y. Saito and K. Maruyama, *J. Chem. Soc. Chem. Commun.* 1326 (1982).

– 6 –

Arylsilanes

6.1. PREPARATION

The preparation of aryl- and heteroaryltrimethylsilanes has been comprehensively reviewed (1). In practice, these silanes are prepared most frequently by the quenching of a suitable organometallic derivative with TMSCl:

$$\text{ArX} \longrightarrow \text{ArM} \xrightarrow{\text{TMSCl}} \text{ArSiMe}_3$$

A subsidiary approach involves nuclear modification of the arylsilanes so obtained. The requisite organometallics can be prepared from aryl halides, or by deprotonation of a suitably activated (e.g. methoxy-substituted) arene. A more specialized route involves cycloaddition between alkynylsilanes and diynes.

6.1.1. Grignard Route

General procedure (2)

To a stirred mixture of Mg (0.25 g atom), HMPA (CAUTION—CANCER SUSPECT AGENT) (120 ml) and TMSCl (0.25 mol) was added a few drops of the aryl halide (0.2 mol) in HMPA (50 ml). The mixture was then warmed until foaming commenced, and the reaction initiated by the addition of a few drops of 1,2-dibromoethane. The remaining solution was added over 5 h at 80 °C with stirring, which was continued for a further 24–28 h at this temperature. After cooling, the mixture was added with stirring to ice-cold sodium hydrogen carbonate solution (1 l, 5 g/l), and the continued neutrality of the mixture tested. The mixture was then filtered by suction, the filtrate extracted with ether, and the ethereal extracts were washed with water and dried. Concentration and appropriate purification by distillation or crystallization gave the arylsilanes (60–87%).

39

6.1.2. Lithium–Halogen Exchange

Trimethyl-m-dimethylaminophenylsilane (3)

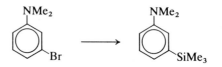

To a suspension of finely cut Li (0.81 g atom) in ether (100 ml) was added, with stirring, *m*-bromodimethylaniline (0.35 mol) in ether (100 ml) over a 2 h period. The reaction mixture was heated under reflux with stirring for a further 0.5 h, and then a solution of TMSCl (0.34 mol) in ether (75 ml) was added over 1 h. The cooled reaction mixture was poured into cold saturated ammonium chloride solution, and the organic layer was washed with brine and dried. Concentration and distillation gave trimethyl-*m*-dimethyl-aminophenylsilane (0.294 mol, 84%), b.p. 109–110 °C/8 mmHg.

6.1.3. *o*-Lithiation

4-Chloro-2-trimethylsilylanisole (4)

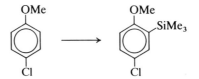

A mixture of *p*-chloroanisole (0.5 mol) and PhLi (0.5 mol, prepared from PhBr and Li in ether) in ether (600 ml) was stirred at ambient temperature for 50 h. A solution of TMSCl (0.5 mol) in ether (50 ml) was added with stirring, which was continued for a further 24 h. The mixture was poured into saturated ammonium chloride solution, the layers were separated, and the ethereal layer was dried. Concentration and fractional distillation gave 4-chloro-2-trimethylsilylanisole, b.p. 84 °C/2 mmHg, which crystallized in the receiver. Three crystallizations from ethanol gave the pure silane (0.3 mol, 60%), m.p. 51–51.5 °C.

6.1.4. Rearrangement of Metallated Aryl Silyl Ethers

2- and 4-Trimethylsilylphenol (5)

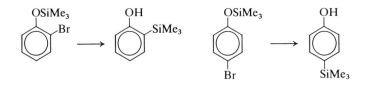

To a solution of 2-(or 4-)bromophenyl trimethylsilyl ether (10 mmol) in THF (50–100 ml), cooled to −78 °C, was added t-BuLi (20 mmol, 15% in pentane) with stirring. After 30 min at −30 °C (or 24 h at +25 °C), the reaction mixture was added to saturated ammonium chloride solution and extracted with ether. Drying, concentration and appropriate purification gave 2-trimethylsilylphenol (7.6 mmol, 76%), b.p. 85 °C/12 mmHg (4-trimethylsilylphenol (8.5 mmol, 85%), m.p. 74 °C).

6.2. REACTIONS

Replacement of an aromatic/heteroaromatic proton with a trialkylsilyl group can confer a variety of synthetic advantages. The silyl moiety can mask a potentially acidic proton, and it can be readily removed by electrophiles, normally resulting in a process of *ipso* desilylation:

For example, all three isomeric aryl acetates (**1**) undergo bromo- and iododesilylation, providing a route (*6*) to radio-halogen-labelled phenols; the free phenols and methyl ethers corresponding to (**1**) proved too reactive, giving products of both substitution and desilylation.

(**1**)

6.2.1. Iododesilylation of Aryltrimethylsilanes

An alternative method (7) uses Ag(I) catalysis, and allows the preparation of, *inter alia*, 4-iodophenylalanine, with the additional potential for radio-labelling:

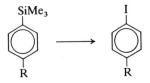

General procedure (7)

A solution of the aryltrimethylsilane (2 mmol) in MeOH (10 ml, purified by being heated under reflux with Mg turnings and a few drops of tetrachloromethane, and distilled immediately prior to use) was cooled to 0 °C with stirring. Silver trifluoroacetate (4.2 mmol) was added, and stirring was continued for 5 min to ensure complete dissolution. Iodine (4 mmol) was added, and stirring was continued at 0 °C for 1 h, at which time g.l.c. analysis indicated completion. Ether (20 ml) was added, the mixture was filtered through a pad of Celite, and the filter cake washed with more ether. The combined organic filtrates were washed with dilute aqueous sodium thiosulphate (2 × 10 ml), dried, and concentrated *in vacuo* to yield the iodoarene (60–85%).

6.2.2. Acylation

2-Methylacetophenone (8)

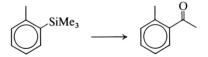

Acetyl chloride (60 mmol) was added dropwise with stirring to an ice-cold mixture of 2-trimethylsilyltoluene (60 mmol) and aluminium trichloride (60 mmol) in carbon disulphide (60 ml). Stirring was continued for 3 h at 0–5 °C, and then the reaction mixture was poured on to ice-cold dilute HCl.

The organic layer was separated, dried and concentrated. Distillation of the residue gave 2-methylacetophenone (88%).

6.2.3. Butoxide-Catalysed Reaction with Aldehydes

Aryltrimethylsilanes also undergo a t-butoxide-catalysed condensation (9) with electrophiles, such as aldehydes:

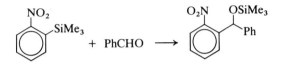

Potassium t-butoxide (0.74 mmol) was added with stirring to a solution of 2-trimethylsilylnitrobenzene (20 mmol) and benzaldehyde (20 mmol) in DMF (25 ml) at ambient temperature. Stirring was continued for 1 h, and the solution was then filtered and concentrated *in vacuo*. Distillation of the residue gave the product (16 mmol, 80%), b.p. 136–138 °C/0.1 mmHg.

REFERENCES

1. *Reviews:* D. Häbich and F. Effenberger, *Synthesis* 841 (1979); L. Birkofer and O. Stuhl, *Top. Curr. Chem.* **88**, 33 (1980); for a good source of more up-to-date material, see F. Effenberger and A. Krebs, *J. Org. Chem.* **49**, 4687 (1984). See also T. H. Chan and I. Fleming, *Synthesis* 761 (1979); Z. Parnes and G. I. Bolestova, *Synthesis* 991 (1984). K. P. C. Vollhardt, *Acc. Chem. Res.* **10**, 1 (1977); R. L. Funk and K. P. C. Vollhardt, *Chem. Soc. Rev.* **9**, 41 (1980). P. Beak and V. Snieckus, *Acc. Chem. Res.* **15**, 306 (1982).
2. F. Effenberger and D. Häbich, *Liebigs Ann. Chem.* 842 (1979).
3. R. E. Benkeser and H. R. Krysiak, *J. Am. Chem. Soc.* **75**, 2421 (1953).
4. C. Eaborn and D. R. M. Walton, *J. Organometal. Chem.* **3**, 169 (1965).
5. G. Simchen and J. Pfletschinger, *Angew. Chem. Int. Edn* **15**, 428 (1976). (For the original study of this rearrangement, on Na salts, including the observation, by crossover, that 2- was intramolecular and 4- intermolecular, and the thermal irreversible rearrangement SiArOH → SiOAr at 250 °C, see J. L. Speier, *J. Am. Chem. Soc.* **74**, 1003 (1952).)
6. D. S. Wilbur, W. E. Stone and K. W. Anderson, *J. Org. Chem.* **48**, 1542 (1983).
7. S. R. Wilson and L. A. Jacob, *J. Org. Chem.* **51**, 4833 (1986).
8. K. Dey, C. Eaborn and D. R. M. Walton, *Organometallics in Chemical Synthesis* **1**, 151 (1970/1971).
9. F. Effenberger and W. Spiegler, *Angew. Chem. Int. Edn* **20**, 265 (1981).

− 7 −

Alkynyl- and Propargylsilanes

Terminal alkynes can be converted readily into alkynylsilanes by reaction of the corresponding alkyne anion or its metalloid equivalent with a suitable chlorosilane (1). The reverse reaction, that of liberation of the alkyne, is quite facile, being effected by several reagent combinations, including hydroxide ion, methanolysis, fluoride anion, silver(I) followed by cyanide anion, and methyl lithium–lithium bromide (2).

$$RC\equiv CH \longleftrightarrow RC\equiv CSiR_3'$$

Terminal silylation of alkynes affords two main benefits: the triple bond is protected against chemical attack, either for steric reasons or because the potentially acidic proton is masked, and, perhaps paradoxically, the bond is activated towards regioselective electrophilic attack under certain conditions.

7.1. PREPARATION

7.1.1. Alkynylsilanes

1-Trimethylsilylethyne (3)

$$HC\equiv CH \longrightarrow HC\equiv CMgBr \longrightarrow HC\equiv CSiMe_3$$

THF (500 ml) was cooled to 0 °C, and saturated with dry ethyne. A solution of EtMgBr (0.8 mol) in THF (400 ml) was added in increments at such a rate that evolution of ethane had ceased before additional Grignard reagent was added. Ethyne was bubbled in throughout this step, and during the initial stages of the next. TMSCl (0.7 mol) was added dropwise, and the solution was stirred at ambient temperature for 5 h. It was then poured into ice-water (500 ml), and the upper layer was separated. Preliminary distillation of this layer separated the more volatile product from the bulk of the THF. The

45

distillate was washed well with water, dried, and redistilled to give trimethylsilylethyne (0.49 mol, 70% based on TMSCl), b.p. 52.5–53.5 °C.

1-Trimethylsilylhex-1-yne (4)

$$\text{n-C}_4\text{H}_9\text{C}{\equiv}\text{CH} \xrightarrow[\text{2. TMSCl}]{\text{1. n-BuLi}} \text{n-C}_4\text{H}_9\text{C}{\equiv}\text{CSiMe}_3$$

A solution of hex-1-yne (0.3 mol) in ether (100 ml) was treated consecutively at −78 °C with a solution of n-butyl lithium (0.306 mol) in hexane and with TMSCl (0.306 mol). The reaction mixture was brought to ambient temperature, stirred for 2 h, and then quenched with ice-water. The layers were separated, and the aqueous layer was re-extracted with pentane. The combined organic extracts were washed with water and brine, and dried. Concentration and distillation gave the alkynylsilane (0.267 mol, 89%), b.p. 71–73 °C/36 mmHg.

7.1.2. Propargylsilanes

Both terminal and internal propargylsilanes are versatile intermediates for the preparation of allylsilanes (Chapter 5).

1-Trimethylsilylhept-2-yne (5)

$$\text{n-C}_4\text{H}_9\text{C}{\equiv}\text{CCH}_3 \xrightarrow[\text{2. TMSCl}]{\text{1. t-BuLi}} \text{n-C}_4\text{H}_9\text{C}{\equiv}\text{CCH}_2\text{SiMe}_3$$

To a solution of t-BuLi (21 mmol, 2.3 M in pentane), cooled to −78 °C, were added sequentially with stirring ether (20 ml), TMEDA (20 mmol), and hept-2-yne (20 mmol). The yellow slurry thus produced was allowed to come to 0 °C, and was stirred at this temperature for a further hour. The yellow solution was then cooled to −78 °C, and treated dropwise with TMSCl (24 mmol). The mixture, on reaching ambient temperature, was poured on to ice-water (20 ml), and the layers were separated. The aqueous layer was extracted with ether (3 × 20 ml), and the combined extracts were washed with aqueous HCl (20 ml, 3 M) and brine, and dried. Concentration and distillation gave the product (16.2 mmol, 81%), b.p. 69–70 °C/10 mmHg.

3-Trimethylsilyloct-1-yne (5,6)

$$\text{n-C}_5\text{H}_{11}\text{CH}_2\text{C}{\equiv}\text{CH} \longrightarrow \text{n-C}_5\text{H}_{11}\text{CH(SiMe}_3)\text{C}{\equiv}\text{CSiMe}_3$$
$$\longrightarrow \text{n-C}_5\text{H}_{11}\text{CH(SiMe}_3)\text{C}{\equiv}\text{CH}$$

To a stirred solution of oct-1-yne (25 mmol) in THF (25 ml) was added dropwise a solution of n-BuLi (26.3 mmol, 1.6 M in hexane), maintaining the reaction temperature during addition below −60 °C. The mixture was stirred at −78 °C for 15 min, then TMSCl (27.5 mmol) was added dropwise. The resulting slurry was allowed to come to ambient temperature, stirred at this temperature for 1 h, and then cooled to −35 °C. To this slurry was added a solution of t-BuLi (27.5 mmol, 2.7 M in pentane), and stirring was continued at −30 °C for 2 h. The reaction mixture was then cooled to −78 °C, and TMSCl (30 mmol) was added dropwise. The mixture was allowed to come to ambient temperature, stirred at this temperature for 1 h, and then poured onto ice-water (25 ml). The layers were separated, and the aqueous layer was extracted with pentane (2 × 25 ml). The combined organic extracts were washed with brine, dried, concentrated, and the residue distilled to give 1,3-bis(trimethylsilyl)oct-1-yne (22.25 mmol, 89%), b.p. 78–80 °C/ 1 mmHg.

To a solution of 1,3-bis(trimethylsilyl)oct-1-yne (10 mmol) in EtOH (20 ml) was added a solution of silver nitrate (15 mmol) in EtOH (18 ml) and water (6 ml) in four equal portions over 45 min with stirring at 0 °C. After being stirred at this temperature for a further 15 min, the white slurry was treated with an aqueous solution of KCN (75 mmol, 9 M) at 0 °C with stirring. On being allowed to come to ambient temperature, the resulting yellow solution was partitioned between hexane (10 ml) and water (10 ml). The layers were separated, and the aqueous layer was extracted with hexane (2 × 10 ml). The combined organic extracts were washed with water and dried. Concentration and distillation gave the product (7.5 mmol, 75%), b.p. 60–61 °C/4 mmHg.

7.2. REACTIONS OF ALKYNYLSILANES

7.2.1. Acylation

$$RC{\equiv}CSiMe_3 \ + \ R'COCl \longrightarrow RC{\equiv}CCOR'$$

General procedure (7)

To a cooled (2–8 °C) solution of AlCl$_3$ in carbon disulphide or nitrobenzene was added dropwise an equimolar mixture of the alkynylsilane and the acid chloride (or anhydride), dissolved in a little of the same solvent. After being stirred for 30 min, the reaction mixture was poured onto dilute sulphuric acid/ice. Normal isolation procedures gave the alkynone (50–90%).

Cyclopentadec-2-ynone (8)

$$Me_3SiC{\equiv}C(CH_2)_{12}COCl \longrightarrow \underset{\underset{(CH_2)_{12}}{\diagdown \diagup}}{O{=}C{-}C{\equiv}C}$$

To a refluxing mixture of freshly sublimed and well ground $AlCl_3$ (3 mmol) and dichloromethane (200 ml) was added a solution of 15-trimethylsilyl-pentadec-14-ynoyl chloride (1 mmol) in dichloromethane (25 ml) over 2.5 h, using a high-dilution apparatus (9). After heating under reflux for a further 0.5 h, the reaction mixture was cooled to 0 °C, and quenched with dilute HCl. Layer separation, concentration, and chromatography on silica gel, eluting with benzene, gave cyclopentadec-2-ynone (0.77 mmol, 77%).

7.2.2. Generation of Formal Alkyne Anions

Phenyl trimethylsilylethyne (Chapter 20) undergoes a fluoride-ion-catalysed addition to aldehydes and ketones. This provides a remarkably mild, relatively non-basic method for the generation of an alkynyl anion or its equivalent.

1-Phenyl-3-trimethylsilyloxydec-1-yne (10)

$$PhC{\equiv}CSiMe_3 \ + \ n{-}C_7H_{15}CHO \ \xrightarrow{\quad F^- \quad} \ n{-}C_7H_{15}CH(OSiMe_3)C{\equiv}CPh$$

A solution of phenyl trimethylsilylethyne (2.3 mmol) and octanal (2 mmol) in THF (2.5 ml) was treated at 0 °C with TBAF (0.06 mmol) for 1.5 h. The reaction mixture was diluted with hexane (25 ml), filtered through Celite, and concentrated. Purification by preparative t.l.c., followed by distillation, gave the product (1.4 mmol, 70%), b.p. 140–142 °C/15 mmHg.

Note. This sequence succeeds with both aldehydes and ketones, but fails with readily enolizable ketones and enones.

7.2.3. Cleavage

$$RC{\equiv}CSiMe_3 \longrightarrow RC{\equiv}CH$$

General procedure (i) (11)

To a solution of the alkynylsilane (60 mmol) in EtOH (125 ml) was added a solution of silver nitrate (160 mmol) in water (60 ml) and EtOH over 25 min at ambient temperature. The reaction temperature rose to 30 °C, and precipitation of the silver acetylide occurred. After complete addition, the mixture was stirred for a further 15 min, and then a solution of KCN (770 mmol) in water (75 ml) was added. Stirring was continued until the precipitate had dissolved. Water (300 ml) was added, and the solution was extracted thoroughly with pentane (4 × 100 ml). The pentane extracts were combined, washed with water, dried and concentrated. The residue was distilled to give the free alkyne (80–90%).

Note. This method is recommended for sensitive substrates that would not tolerate KOH or MeLi, such as enynes.

Alternatively, RC≡CSiMe₃ cleavage can be achieved easily, avoiding the use of TBAF, by employing phase-transfer catalysis; the reaction is complete in 5–10 min, and the conditions are compatible with other nucleophically labiele functional groups such as epoxides.

General procedure (ii) (12)

To a stirred solution of the alkynylsilane (20 mmol) and triethylbenzyl-ammonium chloride (0.7 mmol) in MeCN (15 ml) cooled to 0 °C was added aqueous sodium hydroxide (15 ml, 12 M). After 5–10 min, the mixture was diluted with ether and extracted with ether/dichloromethane. Drying, concentration and suitable purification gave the free alkyne (80–90%).

REFERENCES

1. *Reviews:* T. H. Chan and I. Fleming, *Synthesis* 761 (1979); Z. Parnes and G. I. Bolestova, *Synthesis* 991 (1984).
2. A. B. Holmes, C. L. D. Jennings-White, A. H. Schultess, B. Akinde and D. R. M. Walton, *J. Chem. Soc. Chem. Commun.* 840 (1979); A. B. Holmes and G. E. Jones, *Tetrahedron Lett.* 3111 (1980).
3. C. S. Kraihanzel and M. S. Losee, *J. Organometal. Chem.* **10**, 427 (1967).
4. G. Zweifel and W. Lewis, *J. Org. Chem.* **43**, 2739 (1978).
5. S. Rajagopalan and G. Zweifel, *Synthesis* 111 (1984).
6. For an earlier report, without details, on the selective desilylation of an acetylenic/propargylic disilane, see B. Bennetau, J.-P. Pillot, J. Dunoguès and R. Calas, *J. Chem. Soc. Chem. Commun.* 1094 (1981).
7. L. Birkofer, A. Ritter and H. Uhlenbrauck, *Chem. Ber.* **96**, 3280 (1963).
8. K. Utimoto, M. Tanaka, M. Kitai and H. Nozaki, *Tetrahedron Lett.* 2301 (1978).
9. A. C. Davies, *Chem. Ind. (Lond.)* 203 (1977).
10. I. Kuwajima, E. Nakamura and K. Hashimoto, *Tetrahedron,* **39**, 975 (1983).
11. H. M. Schmidt and J. F. Arens, *Rec. Trav. Chim.* **86**, 1138 (1967).
12. J. Roser and W. Eberbach, *Synth. Commun.* **16**, 983 (1986).

— 8 —

Silyl Anions

Early work on the chemistry of organosilyl anions/anionoids has been thoroughly reviewed (*1*). The most frequently employed preparative routes involve either cleavage of a disilane, when HMPA (CAUTION—CANCER SUSPECT AGENT) is normally required as solvent, or reaction of bulky silyl chlorides with lithium metal.

8.1. TRIMETHYLSILYL LITHIUM (*2,3*)

$$\text{Me}_3\text{Si—SiMe}_3 \ + \ \text{MeLi} \ \xrightarrow[\ 0\,°\text{C, 15 min}\ ]{\text{HMPA}} \ \text{Me}_3\text{SiLi} \ + \ \text{Me}_4\text{Si}$$

A solution of hexamethyldisilane (2.5 mmol) in HMPA (CAUTION— CANCER SUSPECT AGENT) (2 ml) was cooled to 0 °C, and MeLi (2 mmol, 1 M in ether) added *via* a syringe. The resulting deep-red solution was stirred for 15 min to complete formation of the reagent.

This reagent reacts with α,β-unsaturated ketones to give kinetic products of exclusive 1,4-addition (*2*). With cyclic substrates, a strong preference for axial addition is observed, as is a susceptibility to steric hindrance. Transformation into the corresponding silylcuprate species permits conjugate addition to a wider variety of α,β-unsaturated substrates (*3,4*).

8.1.1. Conjugate Addition to $\alpha\beta$-Unsaturated Ketones

trans-3-Trimethylsilyl-2-methylcyclohexanone (2)

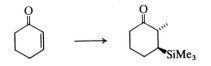

THF (10 ml) was added to the above solution of TMSLi in HMPA, and the mixture immediately cooled to −78 °C. A solution of cyclohex-2-enone (1.5 mmol) in THF (1 ml) was then added dropwise. After stirring for an additional 5 min, methyl iodide (0.5 ml, excess) was added, and the mixture allowed to warm slowly to 0 °C. It was then poured into pentane (50 ml) and washed thoroughly with water (2 × 25 ml). After drying and concentration, the residual oil was distilled to give *trans*-3-trimethylsilyl-2-methylcyclo-hexanone (97%), b.p. 80 °C/1 mmHg (Kugelrohr).

8.2. LITHIUM BIS(TRIMETHYLSILYL)CUPRATE (*3,4*)

$$2Me_3SiLi \; + \; CuCN \; \longrightarrow \; (Me_3Si)_2CuLi$$

A solution of trimethylsilyl lithium (10 mmol) in HMPA (CAUTION—CANCER SUSPECT AGENT) (5 ml) and ether (10 ml, from the MeLi) prepared as above was cooled to 0 °C and diluted with THF (20 ml). Copper(I) cyanide (5 mmol) was added in one portion, and the resulting black mixture was stirred at 0 °C for 20 min.

8.2.1. Conjugate Addition to αβ-Unsaturated Ketones

3,5,5-Trimethyl-3-trimethylsilylcyclohexanone (3)

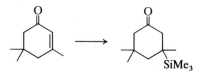

To a stirred solution of the above cuprate reagent (5 mmol), cooled to −23 °C, was added isophorone (3 mmol). The resulting mixture was stirred at −23 °C for 1 h, and then for 30 min at 0 °C. It was quenched by the addition of saturated ammonium chloride solution (1 ml), and then diluted with pentane and filtered through glass wool. The filtrate was washed thoroughly with saturated ammonium chloride solution, dried and concentrated *in vacuo*. Column chromatography gave 3,5,5-trimethyl-3-trimethylsilyl-cyclohexanone (2.7 mmol, 90%).

8.3. TRIMETHYLSILYL SODIUM OR POTASSIUM (*5*)

$$Me_3Si\!-\!SiMe_3 \; + \; 2MH \; \xrightarrow{\; HMPA \;} \; Me_3SiM \; + \; [Me_3SiH] \; \longrightarrow \; 2Me_3SiM$$

To a stirred suspension of NaH or KH (20 mmol) in HMPA (CAUTION—CANCER SUSPECT AGENT) was added hexamethyldisilane (10 mmol) slowly with stirring. A clear yellow-brown solution of trimethylsilyl potassium was obtained immediately; mild heating at 30–40 °C is necessary to prepare trimethylsilylsodium.

8.4. DIMETHYLPHENYLSILYL LITHIUM (6)

$$PhMe_2SiCl + 2Li \longrightarrow PhMe_2SiLi + LiCl$$

Dimethylphenylsilyl chloride (10 mmol) was stirred with lithium (30 mmol, Fison's "shot") in THF (25 ml) at −8 °C for 36 h. The red solution is stable for several weeks at −20 °C.

8.5. LITHIUM BIS(PHENYLDIMETHYLSILYL)CUPRATE (6)

$$2PhMe_2SiLi + CuCN \longrightarrow (PhMe_2Si)_2CuLi$$

Dimethylphenylsilyl lithium (10 mmol) in THF (25 ml) was added to copper(I) cyanide (5 mmol, dried at 120 °C for 10 h) at 0 °C, and the mixture was stirred for 15–20 min.

This reagent is perhaps best thought of as a higher-order mixed cuprate of the type $(R_3Si)_2Cu(CN)Li_2$ (7).

8.5.1. Conjugate Addition to αβ-Unsaturated Esters

(RR,SS)-2-Methyl-3-phenyldimethylsilyl-3-phenylpropionic acid (8)

To a stirred slurry of copper(I) cyanide (110 mmol) in THF (100 ml), cooled to 0 °C, was added a solution of dimethylphenylsilyl lithium (220 mmol, 1.3 M in THF), and the mixture was stirred at 0 °C for a further 30 min. After cooling to −78 °C, a solution of methyl cinnamate (100 mmol) in THF (50 ml) was added, and stirring was continued at −78 °C for 6 h. At this time, iodomethane (300 mmol) (CAUTION—CANCER SUSPECT AGENT) was added, and the mixture allowed to warm to ambient temperature with stirring overnight. It was then partitioned between saturated ammonium chloride solution and pentane, and the pentane layer was separated. The mixture was re-extracted with pentane (4×), and the combined organic extracts were washed with saturated ammonium chloride solution until the aqueous layer was colourless, then washed with brine, dried, and concentrated *in vacuo*.

The crude product was dissolved in MeOH (400 ml), and water (100 ml) and LiOH (313 mmol) were added. The two-phase mixture was heated under reflux for 24 h, and then concentrated *in vacuo* to a volume of 100 ml. Ether (300 ml) and water (200 ml) were added, and the separated organic phase was washed with aqueous LiOH (2 × 100 ml, 6 M). The combined basic extracts were acidified to pH 2 using concentrated sulphuric acid, and then extracted with ether (5 × 100 ml). The combined organic extracts were washed with water and brine, and dried. Concentration *in vacuo* followed by crystallization (pentane : acetone, 5 : 1) gave the acid (59 mmol, 59%) as white prisms, m.p. 98–99 °C.

Note. For oxidative cleavage, with retention of stereochemistry, of such species, see Chapter 9.

8.5.2. Reversible Protection of Enones (4)

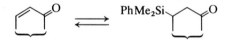

General procedure

Dimethylphenylsilyl lithium (1 mmol, above THF solution) was added to copper(I) iodide (0.5 mmol) at −23 °C, and the mixture was stirred at this temperature for 4 h. The enone (0.75–0.5 mmol) was then added, and stirring was continued at −23 °C for 0.5 h. The mixture was then poured on to ice(25 g)/HCl(5 ml), and extracted with chloroform (3 × 25 ml). The combined extracts were filtered, washed with HCl (25 ml, 3 M), water (25 ml), saturated sodium hydrogen carbonate solution (25 ml) and water (25 ml), and dried. Concentration and purification by preparative t.l.c. (eluting solvent 3 : 7 ether : petrol) gave the β-silylketone (40–99%).

A solution of the β-silylketone (1 mmol) in chloroform (3 ml) was added to a suspension of copper(II) bromide (2 mmol) in boiling ethyl acetate (3 ml). The mixture was heated under reflux for 0.75 h, and then cooled, diluted with carbon tetrachloride (10 ml), filtered, and the precipitate washed with carbon tetrachloride (3×7.5 ml). The combined organic extracts were concentrated, and the product enone was isolated by preparative t.l.c. (43–72%).

Note. With non-cyclic β-silylketones, β-bromoketones are obtained; these undergo facile base-induced elimination to the corresponding enones.

REFERENCES

1. D. D. Davis and C. E. Gray, *Organometal. Chem. Rev.* **6**, 283 (1970).
2. W. C. Still, *J. Org. Chem.* **41**, 3063 (1976); W. C. Still and A. Mitra, *Tetrahedron Lett.* 2659 (1978).
3. I. Fleming and T. W. Newton, *J. Chem. Soc. Perkin Trans. I* 1805 (1984).
4. D. J. Ager, I. Fleming and S. K. Patel, *J. Chem. Soc. Perkin Trans. I* 2520 (1981).
5. R. J. P. Corriu and C. Guerin, *J. Chem. Soc. Chem. Commun.* 168 (1980).
6. I. Fleming, T. W. Newton and F. Roessler, *J. Chem. Soc. Perkin Trans. I* 2527 (1981).
7. B. H. Lipshutz, *Synthesis* 325 (1987).
8. D. C. Parker, Ph.D. thesis, Cambridge University (1985); I. Fleming, personal communication.

– 9 –

Oxidative Cleavage *via* Organofluorosilicates

This excellent method of oxidative cleavage (*1*) of carbon–silicon bonds requires that the silane carry an electronegative substituent (*2*), such as alkoxy or fluoro. Either hydrogen peroxide or mcpba may be used as oxidant, and the alcohol is produced with *retention of configuration* (*3*). Fluoride ion is normally a mandatory additive in what is believed to be a fluoride ion-assisted rearrangement of a silyl peroxide, as shown below:

$$R_3SiF \xrightarrow[\text{oxidant}]{F^-} 3\ ROH$$

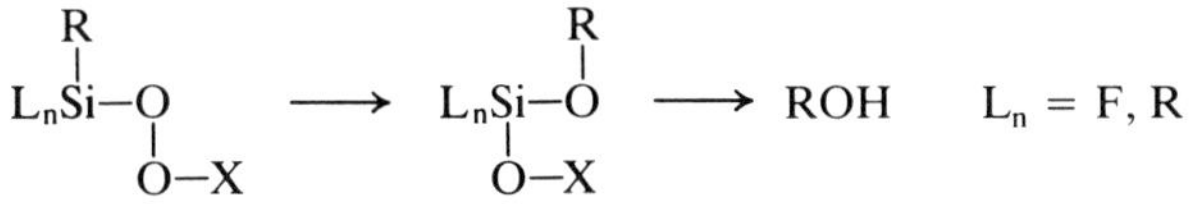

$$\underset{\underset{\displaystyle O-X}{|}}{\overset{\overset{\displaystyle R}{|}}{L_nSi-O}} \longrightarrow \underset{\underset{\displaystyle O-X}{|}}{\overset{\overset{\displaystyle R}{|}}{L_nSi-O}} \longrightarrow ROH \qquad L_n = F, R$$

This possibly inconvenient substitution requirement can be created readily by protiodesilylation of allyldimethylsilyl (*4*) or phenyldimethylsilyl (*5*) moieties with HF equivalents.

9.1. PREPARATION OF POTASSIUM ORGANOPENTAFLUOROSILICATES

$$RSiCl_3 \ + \ \text{excess KF} \longrightarrow K_2[RSiF_5]$$

General procedure (6)

The organotrichlorosilane (100 mmol) was added dropwise to a solution of KF (2.5 mol) in water (220 ml) at 0 °C with vigorous stirring. (The reaction is exothermic, and it is essential to maintain the reaction temperature at 0 °C.) The organopentafluorosilicate began to precipitate immediately as a white solid. After complete addition, the mixture was stirred at ambient temperature for 2 h to overnight. The precipitate was then suction-filtered, and washed successively with water (300 ml), ethanol (100 ml) and ether (100 ml), and then dried in a vacuum desiccator over P_2O_5 (74–98%).

9.2. mcpba CLEAVAGE

Methyl 11-hydroxyundecanoate (7)

$$K_2[MeOOC(CH_2)_{10}SiF_5] \xrightarrow{\text{mcpba}} HO(CH_2)_{10}COOMe$$

$K_2[MeO_2C(CH_2)_{10}SiF_5]$ (prepared by the platinum-catalysed hydrosilylation of methyl 10-undecenoate with trichlorosilane, followed by treatment with potassium fluoride in water) (3 mmol), mcpba (80% pure, 3.6 mmol as active oxygen) and DMF (10 ml) were combined and stirred at room temperature for 6 h, by which time the mixture had become almost homogeneous. After addition of ether (100 ml), the mixture was washed with water (25 ml), aqueous sodium bisulphite (2 × 25 ml, 20%) and saturated sodium hydrogen carbonate solution (2 × 25 ml), and dried. Concentration and distillation gave methyl 11-hydroxyundecanoate (2.3 mmol, 77%).

Such methodology can be seen in the construction of the nucleophilic hydroxymethylating reagents (isopropoxydimethylsilyl)methylmagnesium chloride (**1**) (*8*) and (allyldimethylsilyl)methylmagnesium chloride (**2**) (*4*):

(i-PrO)Me₂SiCH₂MgCl   (CH₂=CHCH₂)Me₂SiCH₂MgCl

(**1**)   (**2**)

9.3. HYDROXYMETHYLATION OF ALDEHYDES AND KETONES

Nonane-1,2-diol (8)

$$\text{n-}C_7H_{15}CHO \longrightarrow \text{n-}C_7H_{15}CH(OH)CH_2OH$$

To a solution of the Grignard reagent (7.2 mmol) prepared from (i-PrO)-Me₂SiCH₂Cl (prepared from ClMe₂SiCH₂Cl by treatment with i-PrOH (1.5 eq) and gaseous ammonia in ether, 80% yield, b.p. 64–66 °C/ 54 mmHg) and magnesium in THF (25 ml) was added n-octanal with stirring at 0 °C. Stirring was continued at this temperature for 1 h, and then saturated ammonium chloride solution was added. The organic layer was separated, and the aqueous layer was extracted with ether. The organic layers were combined, washed with water, dried briefly, and then concentrated under reduced pressure at 0 °C. All operations up to this point were carried out quickly at around 0 °C. To the concentrate were added MeOH (10 ml), THF (10 ml), sodium hydrogen carbonate (4 mmol) and hydrogen peroxide (30%, 3.6 ml, *ca.* 36 mmol), and the mixture was heated under reflux with stirring for 15 h. On cooling, well-ground sodium thiosulphate pentahydrate (3 g) was added, causing a mild exothermic reaction. After being stirred for 30 min, the solution was negative towards starch–iodide. It was then diluted with ether (20 ml), filtered with suction through Celite, and concentrated. The evaporation residue was taken up in ether (50 ml), dried, concentrated and distilled, to give nonane-1,2-diol (82%), b.p. 130–140 °C/15 mmHg.

Note. The related reagent (diisopropoxymethylsilyl)methylmagnesium chloride (*9*) has been used similarly to hydroxymethylate organic halides and mesylates.

9.4. CONJUGATE HYDROXYMETHYLATION OF αβ-UNSATURATED KETONES

3-Hydroxymethyl-3,5,5-trimethylcyclohexanone (4)

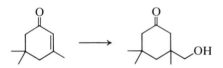

An ethereal solution of (allyldimethylsilyl)methylmagnesium chloride (prepared from allyldimethylsilylmethyl chloride (prepared in turn from ClMe$_2$SiCH$_2$Cl and allylmagnesium chloride in ether in 85% yield, b.p. 63–66 °C/50 mmHg) and magnesium) (1 M, 6 mmol) was added to a suspension of CuI (0.4 mmol) in ether (3 ml) with stirring at 0 °C, followed by a solution of 3,5,5-trimethylcyclohexenone (4 mmol) in ether (10 ml). After stirring the mixture for 1 h at ambient temperature, saturated ammonium chloride solution was added at 0 °C. The ethereal layer was separated, and the aqueous layer re-extracted with ether. The combined organic extracts were washed with water, dried, and concentrated under reduced pressure. To the concentrate were added chloroform (5 ml), KHF$_2$ (8 mmol) and trifluoroacetic acid (12 mmol), and the mixture was heated at 50 °C for 3 h. After concentration, the crude fluorosilane was diluted with MeOH (10 ml) and THF (10 ml), and sodium hydrogen carbonate (20 mmol) and hydrogen peroxide (30%, 72 mmol) were added. The mixture was heated under reflux for 10 h, and then cooled and concentrated (CAUTION—PEROXIDES) under reduced pressure. The residue was taken up in ether (50 ml), dried, concentrated and distilled, to give the hydroxyketone (68%), b.p. 125–135 °C/21 mmHg.

9.5. CONVERSION OF TERMINAL ALKYNES INTO CARBOXYLIC ACIDS

Terminal alkynes can be converted by a process of hydrosilylation followed by oxidative cleavage into carboxylic acids (10). An alternative "basic" cleavage yields the corresponding aldehydes.

n-Octanoic Acid (10)

$$n\text{-}C_6H_{13}C\equiv CH \longrightarrow n\text{-}C_6H_{13}CH\!=\!CHSiMe(OEt)_2$$
$$\longrightarrow n\text{-}C_6H_{13}CH_2COOH$$

To a mixture of 1-octyne (3.01 mmol) and HSiMe(OEt)$_2$ (3.66 mmol) were added a few drops of solution of H$_2$PtCl$_6$.6H$_2$O in i-PrOH (0.1 M), at ambient temperature with stirring. An exothermic reaction ensued, and was complete in 30 min. To the resulting mixture were added successively KHF$_2$ (7.31 mmol), DMF (10 ml) and acetic anhydride (8.8 mmol), followed by hydrogen peroxide (30%, 0.88 ml, *ca.* 8.8 mmol), dropwise at ambient temperature with stirring. An initial exothermic reaction occurred, and the mixture was stirred at ambient temperature overnight. Normal work-up procedures were followed by alkaline extraction, acidification of the alkaline extract, extraction of this with ether, drying of the ethereal extract, concentration and distillation, to give n-octanoic acid (2.04 mmol, 68% yield overall).

9.6. F⁻ INDUCED OXIDATIVE CLEAVAGE OF PHENYLDIMETHYLSILYL GROUPS (11)

(RR,SS)-3-Hydroxy-2-methyl-3-phenylpropionic acid (11)

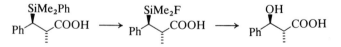

To a solution of (*RR,SS*)-2-methyl-3-phenyldimethylsilyl-3-phenyl-propionic acid (50 mmol) (see Chapter 8) in dichloromethane (50 ml), cooled to 0°C, was added, with stirring, $HBF_4.2Et_2O$ (60 mmol), and the mixture was stirred at ambient temperature overnight. It was then diluted with dichloromethane (200 ml), and washed rapidly with cold water (50 ml) and cold brine (2 × 50 ml), and dried. Concentration *in vacuo* gave the crude fluorosilane, which was used without further purification.

The fluorosilane was dissolved with stirring in peracetic acid (72 ml, 40% solution in AcOH) at 0°C. Triethylamine (53 mmol) was added dropwise, and the mixture was allowed to come to ambient temperature. After being stirred at ambient temperature for 3.25 h, the mixture was poured into aqueous KOH (250 ml, 2 M), and the pH was brought to 11 by the addition of solid KOH pellets. The resulting mixture was extracted with ether (2 × 100 ml), and the aqueous phase was carefully acidified with concentrated sulphuric acid to pH 1. This acidified phase was then extracted with ether (4 × 100 ml), and the combined organic extracts were dried. Concentration *in vacuo* and crystallization (pentane : acetone, 3 : 1) gave the β-hydroxy acid (44.5 mmol, 89%), m.p. 92–94°C.

REFERENCES

1. For earlier work on this process, see R. Müller. *Z. Chem.* **24**, 41 (1984); E. Buncel and A. G. Davies, *J. Chem. Soc.* 1550 (1958).
2. K. Tamao, N. Ishida, T. Tanaka and M. Kumada, *Organometallics* **2**, 1694 (1983).
3. K. Tamao, T. Kakui, M. Akita, T. Iwahara, R. Kanatani, J. Yoshida and M. Kumada, *Tetrahedron* **39**, 983 (1983).
4. K. Tamao and N. Ishida, *Tetrahedron Lett.* **25**, 4249 (1984).
5. I. Fleming, R. Henning and H. Plaut, *J. Chem. Soc. Chem. Commun.* 29 (1984).
6. K. Tamao, J.-I. Yoshida, H. Yamamoto, T. Kakui, H. Matsumoto, M. Takahashi, A. Kurita, M. Murata and M. Kumada, *Organometallics* **1**, 355 (1982).
7. K. Tamao, T. Kakui and M. Kumada, *J. Am. Chem. Soc.* **100**, 2268 (1978).
8. K. Tamao and N. Ishida, *Tetrahedron Lett.* **25**, 4245 (1984).
9. K. Tamao, N. Ishida and M. Kumada, *J. Org. Chem.* **48**, 2120 (1983); M. A. Tius and A. Fauq, *J. Am. Chem. Soc.* **108**, 6389 (1986).
10. K. Tamao, M. Kumada and K. Maeda, *Tetrahedron Lett.* **25**, 321 (1984).
11. D. C. Parker, Ph.D. thesis, Cambridge University (1985); I. Fleming, personal communication; I. Fleming and P. E. J. Sanderson, *Tetrahedron Lett.* **28**, 4229 (1987); for a recent example of use, see W. Oppolzer, R. J. Mills, W. Pachinger and T. Stevenson, *Helv. Chim. Acta* **69**, 1542 (1986).

– 10 –

Peterson Olefination

In a general process, organosilanes that carry a potential leaving group at the β-position can undergo a facile 1,2-elimination reaction (*1*). In those cases where the β-substituent is hydroxyl, the process is known as the silyl-Wittig reaction, or more frequently as Peterson olefination (*2*). Elimination can be induced using either acidic or basic conditions. The simplest application can be seen in the methylenation of carbonyl compounds using the Grignard reagent Me_3SiCH_2MgCl. This can be prepared readily by reaction of Mg with chloromethyltrimethylsilane in either ether (*3*) or THF (*2*), and it reacts in a predictable way with a wide variety of carbonyl compounds. When compared with the Wittig ylide $Ph_3P{=}CH_2$, it is sterically less hindered but more basic.

$$\left\{\!\!\!\searrow\!\!=\!O \longrightarrow \left\{\!\!\!\!\diagup\!\!\!\!\!\!\swarrow\!\!\!\!\!\begin{array}{l}OH\\CH_2SiMe_3\end{array} \longrightarrow \left\{\!\!\!\searrow\!\!=$$

10.1. PREPARATION OF TERMINAL ALKENES

Non-1-ene (2)

$$n\text{-}C_7H_{15}CHO \longrightarrow n\text{-}C_7H_{15}CH(OH)CH_2SiMe_3 \longrightarrow n\text{-}C_7H_{15}CH{=}CH_2$$

Reaction between $TMSCH_2MgCl$ and n-octanal in refluxing THF, followed by isolation by pouring on to ice and saturated ammonium chloride solution and normal extractive procedures gave 2-hydroxynonyltrimethylsilane, b.p. 82–83 °C/1 mmHg.

Base-induced elimination (2). KH (5 g, 40% dispersion in oil, 60 mmol) was washed with hexane. To a stirred slurry of the residue in THF (50 ml) was added the above alcohol (50 mmol); within a few minutes, hydrogen evolution was complete. The mixture was stirred for 6 h at ambient temperature, and then carefully poured on to ice-cold saturated ammonium chloride solution. The layers were separated, and the aqueous layer was extracted with ether (2×50 ml). The combined organic extracts were washed with brine and dried. Careful concentration and distillation gave non-1-ene (35 mmol, 70%), b.p. 143–145 °C/760 mmHg.

Acid-induced elimination (2). To a solution of the above alcohol (40 mmol) in THF (40 ml) was added concentrated sulphuric acid (3 drops), and the mixture was heated at 65 °C for 2 h. After cooling, normal work-up and distillation gave non-1-ene (29 mmol, 73%).

Trimethylsilylmethyl lithium, available as a 1 M solution in pentane (Aldrich), shows similar properties, including high basicity. This basicity can be greatly reduced, and nucleophilicity enhanced, by combining the reagent with anhydrous cerium(III) chloride (4), with the presumed formation of an RCeCl$_2$ species. Addition of aldehydes or ketones to this reagent, followed by addition of TMEDA immediately prior to work-up, gives excellent yields of β-hydroxysilanes (5). Subsequent elimination can be performed under the basic conditions outlined above, or, more advantageously, by using aqueous HF with or without the presence of pyridine:

$$\left\{ \!\! \right\}\!\!=\!\!O \xrightarrow[\text{CeCl}_3]{\text{Me}_3\text{SiCH}_2\text{Li}} \left\{ \!\! \right\}\!\!\begin{array}{c}\text{OH}\\\text{CH}_2\text{SiMe}_3\end{array} \xrightarrow{\text{HF}} \left\{ \!\! \right\}\!\!=$$

General procedure using Me$_3$SiCH$_2$Li/CeCl$_3$ (5)

A flask (25 ml) containing cerium(III) chloride heptahydrate (1.75 mmol) was heated to 140 °C over 1 h at *ca.* 0.1 mmHg. After 1 h at 140 °C, a stirring bar was added, and the contents were stirred for a further 1 h at high vacuum. On cooling to ambient temperature, THF (5 ml) was added, and stirring was continued for 2 h at ambient temperature. The resulting slurry was cooled to −78 °C, and trimethylsilylmethyl lithium (1.5 mmol, 1 M in pentane) added dropwise with vigorous stirring, which was continued at −78 °C for a further 30 min. A solution of the aldehyde or ketone (1 mmol) in THF (1 ml) was added, and stirring was continued at −78 °C for 2–5 h. The reaction mixture was allowed to warm to ambient temperature, and TMEDA (1.75 mmol) was added. After being stirred for 15 min, the mixture was partitioned between dichloromethane (50 ml) and saturated sodium hydrogen carbonate solution (15 ml). The aqueous layer was extracted thoroughly with dichloromethane (3×50 ml), and the combined organic extracts were washed with brine, dried, and concentrated *in vacuo*. Ether (40 ml) was added, and any solid material filtered off. Concentration of the filtrate and purification by flash chromatography gave the pure β-hydroxysilanes.

HF-induced elimination (5). To a solution of aqueous HF (4 drops, 50%) in MeCN (8 ml) was added a solution of the β-hydroxysilane (1 mmol) in MeCN (2 ml), and the mixture was stirred at room temperature until t.l.c. analysis indicated completion (5–20 min). The reaction mixture was then partitioned between pentane (50 ml) and saturated sodium hydrogen carbonate solution (10 ml). The aqueous layer was extracted thoroughly with pentane (3 × 50 ml), and the combined organic extracts were washed with brine and dried. Concentration followed by chromatographic purification gave the product alkenes.

Alternatively, the above process can be performed with pyridine (0.5 ml) also being present in the original HF/MeCN mixture.

10.2. PREPARATION OF GEOMETRICALLY DEFINED ALKENES

The main utility of Peterson olefination lies in the contrasting stereochemical requirements (*6*) for elimination, use of base requiring a *syn* conformation whereas acid conditions demand an *anti* conformation, with complementary geometrical results:

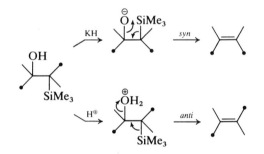

Given a good diastereoisomeric excess in the initial formation of the β-hydroxysilane, either geometrical isomer of the alkene can be obtained.

(E)- and (Z)-Oct-4-ene (6)

These alkene isomers are separately available (*4*) by treatment of *threo*-5-trimethylsilyloctan-4-ol, prepared by reduction of the corresponding ketone with DIBAL in pentane at −120 °C, with base or acid. The preparation of 5-trimethylsilyloctan-4-one itself illustrates three general procedures: the addition of alkyl lithium reagents to vinylsilanes to generate α-lithiosilanes, the preparation of complex β-hydroxysilanes, as diastereoisomeric mixtures, and the oxidation of such compounds to β-ketosilanes

(Chapter 11):

Me₃Si── $\xrightarrow{\text{EtLi}}$ Me₃Si──Li── $\xrightarrow{\text{PrCHO}}$ Me₃Si──OH── $\xrightarrow{\text{CrO}_3.2\text{py}}$ Me₃Si──O
‖   │   │   │   │   │
                          Pr   Pr Pr   Pr Pr

5-Trimethylsilyloctan-4-ol. To a solution of vinyltrimethylsilane (6.77 mmol) in THF (15 ml), cooled to −78 °C, was added a solution of ethyllithium (8.8 mmol, 1.15 M in ether) with stirring. The mixture was stirred at −78 °C for 2 h, warmed to −30 °C over 1 h, and then cooled again to −78 °C. n-Butanal (7.5 mmol) was added, and then the reaction mixture was allowed to come to ambient temperature and stirred for 3 h. It was then partitioned between brine and ether. The layers were separated, and the organic layer was dried, concentrated and the residue distilled (oven temperature 120 °C) to give the β-hydroxysilane (6.3 mmol, 89% based on n-butanal), as a 2 : 1 mixture of *threo* and *erythro* diastereoisomers.

5-Trimethylsilyloctan-4-one. Pyridine (149 mmol) and chromium trioxide (74.5 mmol) were added successively with stirring to dichloromethane (150 ml). The mixture was stirred at ambient temperature for 1 h, and then a solution of the above alcohol (10.6 mmol) in dichloromethane (7 ml) was added dropwise. The reaction mixture was stirred at ambient temperature for 10 min, and then filtered with suction through a pad of Celite. The filtrate was partitioned between saturated ammonium chloride solution and ether, and the layers were separated. The organic extract was washed successively with saturated ammonium chloride solution, aqueous HCl (2 M) and finally with saturated sodium hydrogen carbonate solution, and then dried. Concentration and distillation (oven temperature 150 °C) gave the β-ketosilane (7.45 mmol, 70%).

threo-5-Trimethylsilyloctan-4-ol. In one side of a two-bottomed flask was placed a solution of DIBAL (25.2 mmol, 0.96 M in hexane) and pentane (10 ml). In the other side was placed a solution of the above ketone (8.38 mmol) in pentane (20 ml). The flask was cooled to −120 °C for 1 h, and then tilted to mix the contents of both sides. The mixture was kept at −120 °C for 3 h, and then allowed to come to −20 °C overnight. It was then partitioned between aqueous HCl (2 M) and ether, and the layers were separated. The organic extract was washed with saturated sodium hydrogen carbonate solution, and dried. Concentration and distillation (oven temperature 160 °C) gave the β-hydroxysilane (8.3 mmol, 98%), as a 15 : 1 mixture of *threo* and *erythro* mixture of diastereoisomers.

Base-induced elimination. KH (0.1 g, 50% slurry in oil, 1.25 mmol) was washed with pentane (4 ml). To the residue was added THF (5 ml) and the alcohol (0.378 mmol), and the mixture was stirred for 1 h at ambient temperature. It was then partitioned between ether and saturated ammonium chloride solution, and the ethereal layer was separated, dried and concentrated to give the alkenes as a 95 : 5 mixture of (E) : (Z) isomers, in 96% yield (g.l.c.).

Acid-induced elimination. To a solution of the alcohol (0.632 mmol) in THF (10 ml) was added concentrated sulphuric acid (2 drops), and the mixture was stirred at ambient temperature for 10 h. It was then partitioned between ether and saturated sodium hydrogen carbonate solution. Work-up as above gave the alkenes as an 8 : 92 mixture of (E) : (Z) isomers, in 99% yield (g.l.c.).

Alternatively, to an ice-cold solution of the alcohol (0.325 mmol) in dichloromethane (10 ml) was added boron trifluoride etherate (4 mmol), and the mixture was stirred at 0 °C for 1 h. Work-up as above gave the alkenes as a 6 : 94 mixture of (E) : (Z) isomers, in 99% yield (g.l.c.).

β-Ketosilanes react with alkyl lithiums in a diastereoselective manner (7), the preferred diastereoisomer being the one predicted on the basis of Cram's Rule; acidic or basic treatment provides a stereoselective route to trisubstituted alkenes.

(E)- and (Z)-4-Methyldec-4-ene (7)

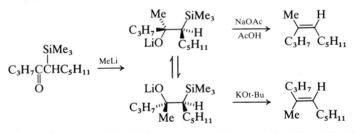

A solution of 5-trimethylsilyldecan-4-one (1 mmol) in THF (5 ml) was treated with MeLi (3 mmol, 0.85 M in ether) at −78 °C with stirring, and the mixture was stirred at ambient temperature overnight. Potassium t-butoxide (9 mmol) was added, and the mixture heated under reflux for 1 h, to give, after work-up, the alkenes as a 91 : 9 mixture of (E) : (Z) isomers, in 74% yield.

Alternatively, treatment of the original reaction mixture (MeLi, −78 °C for 15 min, ambient for 1 h) with AcOH (10 ml) saturated with NaOAc at

−15 °C for 0.5 h with stirring gave, after work-up, the alkenes as a 12 : 88 mixture of (E) : (Z) isomers, in 69% yield.

Until recently, the main route to simple α-lithiosilanes was addition of organolithiums to vinylsilanes, with all the obvious limitations involved. However, reductive lithiation (8) of phenylthioacetals using lithium 1-dimethylaminonaphthalenide (LDMAN) makes the following general process of considerable value:

Lithium 1-(dimethylamino)naphthalenide (LDMAN) (8)

To a flame-dried, argon-purged flask, equipped with a glass-coated stirring bar, was added THF (10 ml) and lithium ribbon (5.8 mg atom). The mixture was cooled to between −45 and −55 °C (hexan-1-ol/dry ice), and 1-(dimethylamino)naphthalene (5.1 mmol) was added slowly. The dark-green colour of the radical anion appeared within 10 min, and formation of LDMAN was complete after 3.5 h of rapid stirring.

Note. This procedure gives a 0.5 M solution of LDMAN, which should be used immediately.

2-(Phenylthio)-2-(trimethylsilyl)propane

A solution of 2,2-bis(phenylthio)propane (38.2 mmol) in THF (50 ml) was added to a solution of LDMAN (92 mmol) in THF (180 ml) at −78 °C, and the resulting mixture was stirred for 15 min. TMSCl (44.3 mmol) was added, and within 1 min the reaction was quenched with excess water at −78 °C. The solvent was removed *in vacuo,* and the residue was taken up in ether. The ethereal solution was washed with aqueous NaOH (2 ×, 1 M), dilute sulphuric acid (2 ×, 1 M), and with saturated sodium hydrogen carbonate solution. After drying and concentration, distillation gave the silane (35.1 mmol, 92%), b.p. 70–80 °C/0.6 mmHg.

1-Cyclohexyl-2-methylpropene

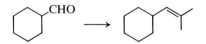

A solution of the above silane (0.91 mmol) in THF (10 ml) was treated with a solution of LDMAN (2.2 mmol) in THF (4.5 ml) at $-78\,°C$ for 15 min. Cyclohexanecarboxaldehyde (0.91 mmol) was added, and within 1 min, the reaction was treated as above. Column chromatography on silica gel gave the alcohol adduct. A solution of this alcohol (1 mmol) in diglyme (2 ml) was treated with KH (excess, hexane-washed) until there was no further hydrogen evolution, and the reaction mixture was warmed to $90\,°C$. After 2 h, the cooled mixture was poured on to ether/ice-water. Separation, drying, concentration and finally purification by column chromatography gave the alkene (0.82 mmol, 92%).

10.3. PREPARATION OF ALLENES

Aldehydes and ketones can be converted into terminal allenes by reaction with α-lithiovinylsilanes followed by elimination (9).

Trideca-1,2-diene (9)

$$n\text{-}C_{10}H_{23}CHO \xrightarrow{\ CH_2=C(Li)SiPh_3\ } n\text{-}C_{10}H_{23}CH(OH)\underset{SiPh_3}{CH}=CH_2$$

$$\xrightarrow[\text{2. Et}_4\text{NF}]{\text{1. SOCl}_2} n\text{-}C_{10}H_{23}CH=C=CH_2$$

To a solution of α-bromovinyl(triphenyl)silane (24 mmol) in ether (60 ml) was added n-BuLi (24 mmol, 1 M in hexane) at $-24\,°C$, and the mixture was stirred for 1.5 h. Undecanal (24 mmol) was added, stirring was continued at $-24\,°C$ for 1 h, and then the reaction was stirred at ambient temperature overnight. It was poured into dilute HCl (50 ml, 3 M), and the organic phase was separated, dried and concentrated, to give the crude alcohol. This was dissolved in tetrachloromethane (25 ml), and treated with thionyl chloride (36 mmol). After being stirred for 2 h, the reaction mixture was concentrated to dryness to give the crude chloride. This in turn was dissolved in DMSO (110 ml), and tetraethylammonium fluoride (28.8 mmol) was added. After being stirred at ambient temperature for 2 h, the mixture was partitioned between ether and water. The separated organic phase was dried and concentrated to give the crude allene, which was purified by distillation (10.6 mmol, 44% yield based on undecanal), b.p. 63–64 °C/0.1 mmHg.

10.4. ALKENE INVERSION *VIA* EPOXIDES

The strict geometrical requirements for elimination can be put to further use, as illustrated by elegant procedures for the geometrical isomerization of alkenes. Trimethylsilyl potassium (*10*) and phenyldimethylsilyl lithium (*11*) both effect smooth conversion of oxiranes into alkenes, nucleophilic ring opening being followed by bond rotation and spontaneous *syn β*-elimination:

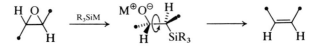

(Z)-Oct-4-ene from (E)*-oct-4-ene oxide (10)*

To a solution of potassium methoxide (0.2 mmol) in HMPA (10 ml) (CAUTION—CANCER SUSPECT AGENT), heated to 65 °C, was added (*E*)-oct-4-ene oxide (1.2 mmol), then hexamethyldisilane (1.8 mmol) in HMPA (5 ml), and the yellow mixture was stirred at 65 °C for 3 h. The cooled mixture was poured onto saturated brine, and extracted with pentane (2 × 25 ml). The combined organic extracts were dried and concentrated, to give oct-4-ene (1.15 mmol, 96%, 99 : 1 (*Z*) : (*E*) by g.l.c.).

(Z)-Stilbene from (E)*-stilbene oxide (11)*

To (*E*)-stilbene oxide (25 mmol) in THF (35 ml) was added freshly prepared dimethylphenylsilyl lithium (25.35 mmol, 1.3 M in THF) dropwise at ambient temperature. The solution was stirred at ambient temperature for 4 h, and then poured into saturated ammonium chloride solution (15 ml), diluted with ether, and the separated organic layer was dried and concentrated. The crude product (97 : 3 (*Z*) : (*E*), g.l.c.) was purified by chromatography on silica gel to give (*Z*)-stilbene (18.75 mmol, 75%).

10.5. PREPARATION OF 3-ALKYLIDENEAZETIDIN-2-ONES

β-Lactams (azetidin-2-ones) can be employed usefully (*12*) in this alkene-forming procedure:

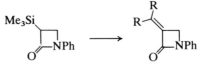

3-Alkylidene-1-phenylazetidin-2-ones (12)

To a solution of LDA (from diisopropylamine (5.6 mmol) and n-BuLi (5.6 mmol, 1.1 M in hexane)) in THF (70 ml) was added a solution of 1-phenylazetidin-2-one (5.1 mmol) in THF (15 ml) at −78 °C. After 3 min, TMSCl (5.6 mmol) was added, and stirring was continued at −78 °C for 5 min. The mixture was poured on to saturated ammonium chloride solution, and extracted with chloroform. The organic extract was washed with water, dried and concentrated, to give the 3-trimethylsilyl β-lactam (4.8 mmol, 95%), which was used directly. To a solution of LDA (from diisopropylamine (5.3 mmol) and n-BuLi (5.3 mmol, 1.1 M in hexane)) in THF (10 ml) was added a solution of the silylated β-lactam (4.8 mmol) in THF (15 ml) at −78 °C. After 3 min, the ketone (5.2 mmol) was added, and the mixture was allowed to stand at −78 °C for 10 min. It was then poured into saturated ammonium chloride solution, stirred at ambient temperature for 5–10 min, and extracted with chloroform. The organic extract was washed with water, dried and concentrated. The alkenes were purified by crystallization (36–93%).

10.6. PREPARATION OF αβ-UNSATURATED ESTERS

Both aldehydes and ketones react with the anion of ethyl trimethylsilyl-acetate to produce αβ-unsaturated esters in an alternative (*13*) to the Reformatsky reaction:

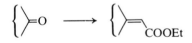

General procedure (13)

A solution of dicyclohexylamine (2 mmol) in THF (10 ml) was treated with n-BuLi (2.025 mmol, 1.5 M in hexane) at −78 °C. After stirring for 15 min at −78 °C, a solution of ethyl trimethylsilylacetate (2 mmol, Chapter 16) in THF (1 ml) was added dropwise, and the resulting solution was stirred for 10 min at −78 °C.

With aldehydes. To a solution of the anion (1.2 mmol) in THF was added dropwise a solution of the aldehyde (1 mmol) in THF (1 ml) at −78 °C. The reaction mixture was stirred at −78 °C for 1 h, −25 °C for 1 h, and +25 °C for 1 h. Finely ground NaHSO$_4$.H$_2$O (0.22 g) was added, and the mixture was stirred for 10 min. The solids were filtered off, water was added, and the mixture was extracted with ethyl acetate (3×5 ml). The combined organic extracts were dried and concentrated, and the product was purified by chromatography.

With ketones. The procedure is identical, but molar ratios are anion : ketone 2 : 1, $NaHSO_4.H_2O$ 0.4 g.

Notes. (*a*) Variable (*Z*) : (*E*) ratios with PhCHO, nonanal 1 : 1.

(*b*) Good results with cyclopentanone, but reaction must be quenched at $-25\,°C$, otherwise up to 10% of the endocyclic alkene isomer is produced.

10.7. USE OF 2-TRIMETHYLSILYLETHANOL

An extension of this general elimination process can be seen in the use of 2-trimethylsilylethanol and its derivatives as protecting/masking groups for molecules with labile hydrogens:

$$R\underset{O}{\overset{O}{\|}}O{\sim}SiMe_3 \xrightarrow{F^{\ominus}} RCOO^{\ominus} + CH_2{=}CH_2 + Me_3SiF$$

The parent alcohol can be obtained in a straightforward manner from chloromethyltrimethylsilane (Chapter 16). The protection afforded is naturally related to the functionality involved, and liberation is achieved using either fluoride ion or a Lewis acid. For example, 2-trimethylsilylethyl esters (Chapter 16) are stable to a wide variety of conditions such as those used in peptide synthesis, but are readily cleaved by fluoride ion (*14,15*).

Variants on this theme can be seen in methods for the protection of amines (*16*), alcohols (*17*), including sugar hemiacetals (*18*), and pyrrole (*19*).

10.8. STEREOSELECTIVE ROUTES TO CONJUGATED ENYNES

Procedure (i) (20)

$$TBDMSCH_2C{\equiv}CSiMe_3 + RCHO \xrightarrow[\text{2. } MgBr_2]{\text{1. } t\text{-BuLi}} (Z)\text{-}RCH{=}CHC{\equiv}CSiMe_3$$

To a solution of $TBDMSCH_2C{\equiv}CSiMe_3$ (2 mmol, preparation given, but not detailed) in THF (6 ml), cooled to $-78\,°C$, was added dropwise, with stirring, a solution of t-BuLi in pentane (2 mmol, 1.8 M). After 1 h at $-78\,°C$, a freshly prepared solution of $MgBr_2$ (2.2 mmol, from Mg and

$BrCH_2CH_2Br$) in ether (6 ml) was added, and stirring was continued for a further 15 min. The carbonyl compound was then added, and after 5 min the solution was warmed slowly to 50 °C and stirred at this temperature for 2–3 h to complete elimination. After cooling, water was added, the solution was extracted repeatedly with ether, and the combined extracts were washed with dilute aqueous HCl and water, and dried. Concentration followed by chromatography on silica gel gave the product (65–90%).

Notes. (*a*) (Z) : (E) 30 : 1 to >50 : 1.
(*b*) Benzaldehyde and cinnamaldehyde were less successful.

Procedure (ii) (21)

$$TIPSCH_2C{\equiv}CTIPS \ + \ RCHO \quad
\begin{array}{l}
\xrightarrow[\text{THF, }-78\,°C]{\text{n-BuLi}} (Z)\text{-}RCH{=}CHC{\equiv}CTIPS \\[2ex]
\xrightarrow[\text{THF/HMPA, }-78\,°C]{\text{n-BuLi}} (E)\text{-}RCH{=}CHC{\equiv}CTIPS
\end{array}$$

$TIPSCH_2C{\equiv}CTIPS$ (preparation given, but not detailed) was metallated (*21*) by treatment with n-BuLi in hexane (1 eq.) in THF at -20 °C for 15 min. Addition of cyclohexane carboxaldehyde (1 eq.) to the anion in THF at -78 °C was followed by gradual warming to ambient temperature over 6 h. Normal extraction procedures followed by chromatography on silica gel afforded the (Z)-enyne, 71%, >20 : 1 (Z) : (E).

Note. Heptanal (57%) and 2,2-dimethylpropanal (79%) give similar stereoselectivity.

Alternatively (*21*), when the same procedure was followed using THF containing 5 eq. HMPA (CAUTION—CANCER SUSPECT AGENT) (relative to the aldehyde) at -78 °C for 15–20 s, followed by quenching at -78 °C and isolation, the (E)-enynes were obtained, (E) : (Z) 20 : 1 to 10 : 1, 60–64%.

10.9. PREPARATION OF N-TRIMETHYLSILYLALDIMINES

Non-enolizable aldehydes are transformed into N-trimethylsilylaldimines on treatment with lithium hexamethyldisilazide (*22*); such imines provide valuable routes to N-unsubstituted β-lactams:

$$RCH{=}O \ + \ LiN(SiMe_3)_2 \xrightarrow{\text{TMSCl}} RCH{=}NSiMe_3 \ + \ LiCl \ + \ (Me_3Si)_2O$$

General procedure (22)

To HMDS (22 mmol) was added n-BuLi (20 mmol, 2.5 M in hexane) over 5 min at ambient temperature. After being stirred for 15 min, the solution was cooled to 0 °C, and THF (36 ml) was added. After a further 20 min at 0 °C, the freshly distilled aldehyde (20 mmol) was added over 7 min. The resulting solution was stirred for 30 min at 0 °C, and then TMSCl (20 mmol) was added in one portion. After a further 40 min, the reaction flask was transferred directly to a rotary evaporator, and the solvent was removed under reduced pressure. The residue was diluted with pentane (15 ml) and filtered rapidly through Celite. Concentration and distillation under reduced pressure (Kugelrohr) gave the pure *N*-trimethylsilylaldimine (76–95%).

REFERENCES

1. D. J. Peterson, *Organometal. Chem. Rev. A* **7**, 295 (1972); P. F. Hudrlik, *J. Organometal. Chem. Library* **1**, 127 (1976); T. H. Chan, *Accts. Chem. Res.* **10**, 442 (1977); D. J. Ager, *Synthesis* 384 (1984).
2. D. J. Peterson, *J. Org. Chem.* **33**, 780 (1968).
3. C. R. Hauser and C. R. Hance, *J. Am. Chem. Soc.* **74**, 5091 (1952).
4. T. Imamoto, T. Kusamoto, Y. Tawarayama, Y. Sugiura, T. Mita, Y. Hatanaka and M. Yokoyama, *J. Org. Chem.* **49**, 3904 (1984).
5. C. R. Johnson and B. D. Tait, *J. Org. Chem.* **52**, 281 (1987).
6. P. F. Hudrlik and D. Peterson, *J. Am. Chem. Soc.* **97**, 1464 (1975).
7. K. Utimoto, M. Obayashi and H. Nozaki, *J. Org. Chem.* **41**, 2940 (1976).
8. T. Cohen, J. P. Sherbine, J. R. Matz, R. R. Hutchins, B. M. McHenry and P. R. Willey, *J. Am. Chem. Soc.* **106**, 3245 (1984); see also D. J. Ager, *J. Chem. Soc. Perkin Trans. I* 183 (1986).
9. T. H. Chan and W. Mychajlowskij, *Tetrahedron Lett.* 171 (1974); T. H. Chan, W. Mychajlowskij, B. S. Ong and D. N. Harpp, *J. Org. Chem.* **43**, 1526 (1978).
10. P. B. Dervan and M. A. Shippey, *J. Am. Chem. Soc.* **98**, 1265 (1976).
11. M. T. Reetz and M. Plachky, *Synthesis* 199 (1976).
12. S. Kano, T. Ebata, K. Funaki and S. Shibuya, *Synthesis* 746 (1978); see also H. Fritz, P. Sutter and C. D. Weis, *J. Org. Chem.* **51**, 558 (1986).
13. H. Taguchi, K. Shimoji, H. Yamamoto and H. Nozaki, *Bull. Chem. Soc. Jpn* **47**, 2529 (1974).
14. P. Sieber, *Helv. Chim. Acta* **60**, 2711 (1977); H. Gerlach, *Helv. Chim. Acta* **60**, 3039 (1977).
15. See for example D. T. W. Chu, J. E. Hengeveld and D. Lester, *Tetrahedron Lett.* **24**, 139 (1983); A. B. Smith and D. Boschelli, *J. Org. Chem.* **48**, 1217 (1983); E. W. Logusch, *Tetrahedron Lett.* **25**, 4195 (1984).
16. L. A. Carpino, J.-H. Tsao, H. Ringsdorf, E. Fell and G. Hettrich, *J. Chem. Soc. Chem. Commun.* 358 (1978); A. I. Meyers, D. L. Comins, D. L. Roland, R. Henning and K. Shimuzu, *J. Am. Chem. Soc.* **101**, 7104 (1979); S. Björkman and J. Chattopadhyaya, *Chem. Scripta* **20**, 201 (1982); R. E. Shute and D. H. Rich, *Synthesis* 346 (1987).
17. B. H. Lipshutz and J. J. Pegram, *Tetrahedron Lett.* **21**, 3343 (1980); C. Gioeli, N. Balgovin, S. Josephson and J. B. Chattopadhyaya, *Tetrahedron Lett.* **22**, 969 (1981).

18. B. H. Lipshutz, J. J. Pegram and M. C. Morey, *Tetrahedron Lett.* **22,** 4603 (1981).
19. M. P. Edwards, S. V. Ley, S. G. Lister and B. D. Palmer, *J. Chem. Soc. Chem. Commun.* 630 (1983); J. M. Muchowski and D. R. Solar, *J. Org. Chem.* **49,** 203 (1984).
20. Y. Yamakodo, M. Ishiguro, N. Ikeda and H. Yamamoto, *J. Am. Chem. Soc.* **103,** 5568 (1981).
21. E. J. Corey and Ch. Rücker, *Tetrahedron Lett.* **23,** 719 (1982).
22. D. McGarry, Ph.D. thesis, University of Glasgow (1985); E. W. Colvin and D. McGarry, *J. Chem. Soc. Chem. Commun.* 540 (1985); see also D. J. Hart, K. Kanai, D. G. Thomas and T.-K. Yang, *J. Org. Chem.* **48,** 289 (1983); D.-C. Ha, D. J. Hart and T.-K. Yang, *J. Am. Chem. Soc.* **106,** 4819 (1984).

– 11 –

β-Ketosilanes

β-Ketosilanes undergo a facile rearrangement to silyl enol ethers (Chapter 15), and are also excellent substrates, after carbonyl reduction, for alkene formation by Peterson olefination (Chapter 10). Such species have, until recently, been somewhat difficult to obtain. Several instances of 1,3 O → C rearrangements have now been discovered that hold significant promise as routes to β-ketosilanes of various structural types. α-Selenocyclohexanones, when converted into hindered silyl enol ethers, undergo reductive cleavage (*1*) of the seleno moiety when treated with LDMAN. The resulting α-lithio silyl enol ethers rearrange rapidly to the silyl enolates, and thence to β-ketosilanes:

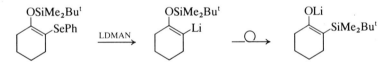

Direct treatment of TIPS enol ethers of a variety of cyclic and acyclic ketones with the strong-base combination of n-BuLi/KO-t-Bu leads to β-ketosilanes (*2*) after aqueous work-up. In contrast with the earlier method, this rearrangement appears to proceed through allylic, rather than vinylic, metallation, since enol ethers lacking an allylic α-proton are unreactive.

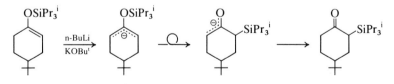

11.1. PREPARATION FROM α-BROMOKETONES

α-Bromoketones, on conversion to the corresponding silyl enol ethers, are transformed into β-ketosilanes (*3*) on treatment with 2 equivalents of

77

n-BuLi:

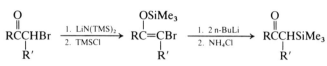

General procedure

The α-bromoketone (non-enolizable on the α' side) was treated with lithium hexamethyldisilazide (1 eq.) at −78 °C in THF, and then TMSCl (1 eq.) was added and the solution allowed to come to ambient temperature. It was then re-cooled to −78 °C, and n-BuLi (2 eq.) was added, and the solution was allowed to come to ambient temperature. The mixture was poured into saturated ammonium chloride solution. Normal work-up and distillation gave the β-ketosilane (50–80%).

Recently it has been shown (*4*) that αβ-dihydroxysilanes, particularly t-butyldimethyl species, undergo an acid-catalysed silapinacol rearrangement to β-aldehydo- and β-ketosilanes, in most respectable yields. The implications of this rearrangement on the acid-catalysed rearrangement of αβ-epoxysilanes to carbonyl compounds are discussed in Chapter 4.

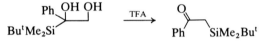

11.2. PREPARATION BY OXIDATION OF β-HYDROXYSILANES

A generally applicable route to β-ketosilanes involves oxidation of β-hydroxysilanes (see also Chapter 10).

Trimethylsilylmethyl cyclohexyl ketone (5)

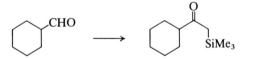

To a solution of trimethylsilylmethyl lithium (from chloromethyltrimethylsilane (15.8 mmol) and lithium dispersion (196 mmol)) in ether (30 ml) was added cyclohexane carboxaldehyde (14.2 mmol) at ambient temperature, with stirring. After a further 10 min at ambient temperature, the solution was filtered, and partitioned between ether (100 ml) and saturated ammonium chloride solution (50 ml). Separation of the ethereal extract, followed by drying, concentration and distillation, gave the adduct alcohol (12.2 mmol, 86%), b.p. 120 °C/0.5 mmHg.

The alcohol (11.9 mmol) was added with stirring to a solution of chromium trioxide (69.4 mmol) and pyridine (133 mmol) in dichloromethane (150 ml) at ambient temperature. After a further 15 min, the solution was filtered, washed with saturated sodium hydrogen carbonate solution, dilute HCl and brine, and dried. Concentration and distillation gave the β-ketosilane (7.4 mmol, 62%), b.p. 120 °C/0.5 mmHg.

Alternatively, secondary and tertiary carboxylic acid methyl or ethyl esters react (6) with two equivalents of trimethylsilylmethyl lithium (prepared in pentane) to give β-ketosilanes in good (80–96%) yield. Primary esters also give β-ketosilanes, but in lower (45%) yield.

$$RCOOMe \ + \ 2LiCH_2SiMe_3 \ \longrightarrow \ RCOCH_2SiMe_3$$

REFERENCES

1. I. Kuwajima and R. Takeda, *Tetrahedron Lett.* **22**, 2381 (1981).
2. E. J. Corey and Ch. Rücker, *Tetrahedron Lett.* **25**, 4345 (1984).
3. P. Sampson and D. F. Weimer, *J. Chem. Soc. Chem. Commun.* 1746 (1985); see also C. J. Kowalski, M. L. O'Dowd, M. C. Burke and K. W. Fields, *J. Am. Chem. Soc.* **102**, 5411 (1980).
4. R. F. Cunico, *Tetrahedron Lett.* **27**, 4269 (1986).
5. R. A. Ruden and B. L. Gaffney, *Synthetic Commun.* **5**, 15 (1975).
6. M. Demuth, *Helv. Chim. Acta* **61**, 3136 (1978).

– 12 –

Acylsilanes

Acylsilanes are most useful synthetic intermediates (*1*), providing, *inter alia*, controlled routes to silyl enol ethers. They are relatively unreactive towards nucleophilic reagents for both steric and electronic reasons.

12.1. PREPARATION

12.1.1. From 1,3-Dithianes

1-Trimethylsilylethanal (2,3)

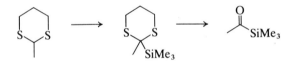

To a solution of 2-methyl-1,3-dithiane (34.1 mmol) in THF (150 ml), cooled to −30 °C, was added n-BuLi (34.1 mmol, 1.5 M in hexane) dropwise (3–5 ml/min). The resulting solution was stirred at −30 to −20 °C for 1.5 h, and then TMSCl (37.1 mmol) was added dropwise. After 2.5 h at −25 °C, water (15 ml) was added, then most of the THF was removed *in vacuo*. Water and pentane were added, the layers were separated, and the aqueous layer was extracted thoroughly with pentane. The combined organic layers were washed with water, aqueous KOH (10%) and water, and dried. Concentration and distillation gave 2-methyl-2-trimethylsilyl-1,3-dithiane (26.8 mmol, 78%), b.p. 102 °C/9.5 mmHg.

All of the following operations were carried out with the exclusion of light. A mixture of 2-methyl-2-trimethylsilyl-1,3-dithiane (121 mmol), mercury(II) chloride (266 mmol), and mercury(II) oxide (181 mmol) in methanol (225 ml) and water (25 ml) was stirred vigorously and heated under reflux for 2 h. It was then cooled and filtered, and the precipitate was washed with boiling methanol. The combined extracts were diluted with water (750 ml), and extracted with pentane (5×). The combined pentane extracts were washed with saturated aqueous ammonium acetate and water, and dried. Concentration and distillation yielded the acylsilane (30 mmol, 25%), b.p. 115 °C/760 mmHg.

Note. Higher hydrolysis yields can be attained using aqueous DMSO and mercury(II) chloride and cadmium carbonate (*3*).

12.1.2. From Thiocarboxylic Acid *S*-Esters

$$R'R''CHCOSR \xrightarrow[\text{2. TMSCl}]{\text{1. LDA}} R'R''C{=}\underset{|}{C}SR \text{ (OSiMe}_3)$$

$$\downarrow Na|TMSCl$$

$$R'R''C{=}\underset{|}{C}SiMe_3 \text{ (OSiMe}_3) \longrightarrow R'R''CH\overset{O}{\overset{\|}{C}}SiMe_3$$

General procedure (4)

To a solution of LDA (from diisopropylamine (24.5 mmol) and n-BuLi (24.5 mmol, 1.64 M in hexane)) in THF (50 ml) was added a solution of the appropriate thioester (20 mmol, R = Me or Ph) in THF (10 ml) over 10 min at −78 °C. TMSCl (30 mmol) was then added. After 1 min, the reaction mixture was allowed to warm to ambient temperature, and was stirred at that temperature for a further 2 h. The solvent was removed *in vacuo*, the residue was diluted with hexane (20 ml), and the precipitated solid was filtered off. Concentration and distillation gave the product ketene silyl thioacetal (83–96%). Sodium dispersion (0.08 g atom, 50% dispersion in paraffin, Alfa Ventron) was washed with hexane (3 × 10 ml) and with benzene (1 × 10 ml). It was then mixed with benzene (35 ml), and introduced into the reaction flask using a hypodermic syringe. TMSCl (80 mmol) was added, followed by a solution of the appropriate ketene acetal (10 mmol) in benzene (15 ml) over 10 min. After addition, the mixture was heated under reflux for 2 h, then cooled, filtered, and concentrated *in vacuo*. The residual oil was washed with THF/saturated sodium hydrogen carbonate solution (1 : 1, 10 ml), and extracted with hexane (20 ml). Drying, concentration and distillation gave the acylsilane silyl enol ether (70–94%); an alternative work-up using 6 N HCl gave the acylsilane (50–70%).

12.1.3. From Allyl Alcohols *via* Brook Rearrangement

1-Trimethylsilylprop-2-en-1-one (5)

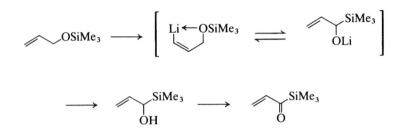

To a solution of allyl alcohol (172.2 mmol) in THF (220 ml), cooled to −78 °C, was added dropwise over 40 min n-BuLi (180.8 mmol, 2.3 M in hexane), with stirring. After 1 h, a solution of TMSCl (180.8 mmol) in THF (20 ml) was added dropwise over 20 min, and stirring was continued for 1.25 h at −78 °C. At this time, t-BuLi (206.6 mmol, 1.6 M in pentane) was added over 40 min, and the resulting solution was stirred for a further 2 h at −78 °C. The cooling bath was removed, and saturated ammonium chloride solution (50 ml) was added. After being stirred for 5 min, the mixture was diluted with water (50 ml) and pentane (300 ml). The organic phase was separated, washed with water (4 × 100 ml) and brine (3 × 50 ml), and dried over sodium sulphate. Filtration followed by careful concentration at atmospheric pressure yielded a residue, which was carefully concentrated further at 15 mmHg to remove remaining volatiles. This gave crude 1-trimethylsilylprop-2-en-1-ol (152 mmol), which was used directly without further purification.

To a solution of oxalyl chloride (171.6 mmol) in dichloromethane (250 ml), cooled to −78 °C, was added dropwise over 45 min a solution of DMSO (372.5 mmol) in dichloromethane, with stirring. After 1 h, a solution of the alcohol (149.2 mmol) in dichloromethane (100 ml) was added dropwise over 1 h, and the resulting solution was stirred for 1 h at −78 °C. At this time, triethylamine (775.8 mmol) was added dropwise over 30 min. After 1 h, the cooling bath was removed, and the reaction mixture was poured into water (150 ml) and pentane (300 ml). The organic phase was washed with dilute HCl (5%, 6×50 ml), water (3×100 ml) and brine (2×100 ml), and dried over sodium sulphate. Concentration and distillation (after the addition of 4,4′-thiobis(2-t-butyl-6-methylphenol) (40 mg)) gave 1-trimethylsilylprop-2-en-1-one (*ca.* 60% overall) as a brilliant yellow oil, b.p. 64–68 °C/30 mmHg.

Notes. (*a*) Rate of metallation with t-BuLi varies from case to case. Lithiation of allyl alcohol trimethylsilyl ether proceeds to completion in 2 h at $-78\,°C$, whereas the corresponding methallyl derivative requires 3.5 h at $-33\,°C$.

(*b*) In the Swern oxidation, the oxalyl chloride must be freshly distilled, and the DMSO carefully dried by distillation from calcium hydride.

12.1.4. From Alkynylsilanes

A convenient route to both saturated and unsaturated acylsilanes lies in the hydroboration–oxidation of alkynylsilanes (Chapter 7). Recent improvements (*6*) to this method involve the use of the borane–dimethyl sulphide complex for hydroboration, and of anhydrous trimethylamine *N*-oxide for the oxidation of the intermediate vinyl boranes.

$$RC{\equiv}CSiMe_3 \xrightarrow[\substack{2.\ Me_3N \to O \\ 3.\ H_2O}]{1.\ BH_3.Me_2S/THF} RCH_2COSiMe_3$$

General Procedure (6)

Borane–dimethyl sulphide solution (73 mmol, 2.74 M in THF) was added dropwise, with stirring, to the neat alkynylsilane (200 mmol), keeping the reaction temperature below 20 °C by use of an ice bath. The mixture was stirred for 2 h at 0–5 °C. At this time, the solution of the trialkenyl borane was diluted with THF (200 ml), and then anhydrous trimethylamine *N*-oxide (240 mmol, dried by azeotropic distillation with toluene) was added over 5 min at ambient temperature. The resultant slurry was heated (bath temperature 75 °C) for 4 h, and then cooled to ambient temperature. The enol borinate thus formed was hydrolysed by the addition of water (100 ml), followed by stirring for 5 min. The layers were separated, and the aqueous phase was extracted with ether (2×100 ml). The combined organic extracts were washed with brine and dried. Concentration and distillation gave the acylsilane (75–91%).

Notes. This sequence fails with t-BuC≡CSiMe₃, owing to steric hindrance, but succeeds well with the sterically less demanding Me₃SiC≡CSiMe₃. General procedure was as above, but with the following amendments.

(*a*) Bis(trimethylsilyl)ethyne (20 mmol) was dissolved in the minimum amount of THF prior to hydroboration.

(*b*) The alkenylborane was oxidized at 25–35 °C for 30 min, and then poured into a mixture of ether (60 ml) and water (30 ml), maintained at 0–5 °C. The mixture was stirred vigorously at this temperature for 15 min, and then worked up as before, to give $Me_3SiCH_2COSiMe_3$ (15 mmol, 75%), b.p. 50 °C/3 mmHg. This acylsilane must be used immediately, since it isomerizes readily to the isomeric silyl enol ether even when stored at low temperatures.

(E)-1-Trimethylsilyl-2,4-dimethylhex-2-en-1-one (R = Me, R' = s-Bu) (6)

$$Me_3SiCH_2COSiMe_3 \xrightarrow{LDA} \overset{Me_3Si}{\underset{H}{>}}C=C\overset{OLi}{\underset{SiMe_3}{<}} \xrightarrow{RX} Me_3SiCH(R)COSiMe_3$$

$$\xrightarrow{LDA} \overset{Me_3Si}{\underset{R}{>}}C=C\overset{OLi}{\underset{SiMe_3}{<}} \xrightarrow{R'CHO} \overset{H}{\underset{R'}{>}}C=C\overset{COSiMe_3}{\underset{R}{<}}$$

To a solution of LDA)prepared from diisopropylamine (22 mmol) in THF (40 ml) and n-BuLi (20 mmol, 2.4 M in hexane)), cooled to −78 °C, was added a solution of the acylsilane (20 mmol) in THF (5 ml). The resulting solution was allowed to warm to 0–5 °C, and stirred at that temperature for 30 min to ensure complete enolate formation. It was then cooled to −25 °C, and a solution of iodomethane (20 mmol) (CAUTION— CANCER SUSPECT AGENT) in THF (10 ml) was added. The mixture temperature. The resulting solution was then added to a solution of LDA (20 mmol, prepared as above), cooled to 0–5 °C. After being stirred at 25 °C for 1 h, the solution of the new enolate was cooled to −78 °C, and a solution of 2-methylbutyraldehyde (22 mmol) in THF (20 ml) was added dropwise over 30 min. The yellow slurry was stirred for a further 15 min at −78 °C, allowed to come to 25 °C, and poured into dilute aqueous HCl (1 M, 25 ml). After thorough extraction with ether, the combined organic phases were washed with brine and dried. Evaporation and distillation gave the (*E*)-*αβ*-unsaturated acylsilane (17 mmol, 85%), b.p. 73–75 °C/4 mmHg.

Notes. (*a*) The alkylating agent must be reactive, i.e. a methyl, ethyl, allyl or benzyl halide.

(*b*) A wide variety of aldehyde structures are suitable; Michael addition does not compete when *αβ*-unsaturated aldehydes are used.

12.2. REACTIONS

Based on the earlier work of Brook (*7*), a number of valuable reactions involving Si—C bond cleavage have been described. Both aromatic and aliphatic acylsilanes have been converted (*8*) into the corresponding aldehydes by treatment with fluoride ion in the presence of a suitable proton source. This transformation seems to proceed, at least in aromatic cases, *via* a formal acyl anion, which can be trapped by a variety of electrophiles:

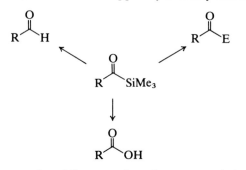

(*E*)-αβ-Unsaturated acylsilanes such as that prepared above undergo (*6*) clean oxidative cleavage to the corresponding carboxylic acids, with retention of double-bond geometry.

Acylsilanes of a variety of substitution patterns have been employed (*9*) in routes to allenyl silyl enol ethers.

REFERENCES

1. *Reviews:* D. J. Ager, *Chem. Soc. Rev.* **11**, 493 (1982); H. J. Reich, M. J. Kelly, R. E. Olson and R. C. Holtan, *Tetrahedron* **39**, 949 (1983).
2. E. J. Corey, D. Seebach and R. Freedman, *J. Am. Chem. Soc.* **89**, 434 (1967).
3. A. G. Brook, J. M. Duff, P. F. Jones and N. R. Davis, *J. Am. Chem. Soc.* **89**, 431 (1967).
4. I. Kuwajima, A. Mori and M. Kato, *Bull. Chem. Soc. Jpn* **53**, 2634 (1980). See also T. Cohen and J. R. Matz, *J. Am. Chem. Soc.* **102**, 6900 (1980) for similar cleavage using LDMAN.
5. R. L. Danheiser, D. M. Fink, K. Okano, Y.-M. Tsai and S. W. Szczepanski, *J. Org. Chem.* **50**, 5393 (1985). For an alternative procedure, see A. Hosomi, H. Hashimoto and H. Sakurai, *J. Organometal. Chem.* **179**, C1 (1979); *J. Org. Chem.* **43**, 2551 (1978). See also W. C. Still, *J. Org. Chem.* **41**, 3063 (1976).
6. J. A. Miller and G. Zweifel, *Synthesis* 288 (1981); J. A. Miller and G. Zweifel, *J. Am. Chem. Soc.* **103**, 6217 (1981).
7. A. G. Brook and A. R. Bassindale, Molecular rearrangements of organosilicon compounds, *Rearrangements in Ground and Excited States*, ed. P. de Mayo, Vol. 2. Academic Press, New York (1980).
8. D. Schinzer and C. H. Heathcock, *Tetrahedron Lett.* **22**, 1881 (1981); A. Degl'Innocenti, S. Pike, D. R. M. Walton, G. Seconi, A. Ricci and M. Fiorenza, *J. Chem. Soc. Chem. Commun.* 1202 (1980); A. Ricci, A. Degl'Innocenti, M. Fiorenza, M. Taddei, M. A. Spartera and D. R. M. Walton, *Tetrahedron Lett.* **23**, 577 (1982).
9. H. J. Reich, E. K. Eisenhart, R. E. Olson and M. J. Kelly, *J. Am. Chem. Soc.* **108**, 7791 (1986).

– 13 –

Aminosilanes

The commercially available dichlorosilane, 1,2-bis(chlorodimethylsilyl)-
ethane, converts primary amines (*1*) into "stabase" derivatives:

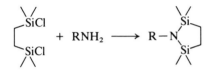

This protecting group is stable to n-BuLi (THF, −25 °C), s-BuLi (Et$_2$O,
−25 °C), LDA, saturated ammonium chloride solution, water, MeOH,
sodium hydrogen carbonate solution (2 M), pyridinium dichromate/
dichloromethane, and, surprisingly, KF.2H$_2$O/THF/H$_2$O. It is unstable to
0.1 M HCl, 1 M KOH, 75% aqueous AcOH, NaBH$_4$, and pyridinium
chlorochromate/dichloromethane.

13.1. PROTECTION OF PRIMARY AMINES

13.1.1. Using 1,2-Bis(chlorodimethylsilyl)ethane

Ethyl glycinate (1)

A solution of 1,2-bis(chlorodimethylsilyl)ethane (8 mmol) in
dichloromethane (3 ml) was added to a stirred solution of ethyl glycinate
(8 mmol) and triethylamine (16 mmol) in dichloromethane (5 ml). The
mixture was stirred at ambient temperature for 2 h, and then it was poured
on to aqueous sodium hydrogen phosphate (5 ml). Separation, drying,
concentration and distillation gave the protected glycinate (7.36 mmol,
92%), b.p. 80–82 °C/0.25 mmHg.

Note. For primary amines in the pK_a range 10–11, triethylamine (2 eq.)
suffices as base; for less-basic amines, n-BuLi (2 eq.) must be used.

13.1.2. Using t-Butylchlorodiphenylsilane

Alternatively, primary amines can be protected (2) as their mono-t-butyl-diphenylsilyl derivatives:

$$RNH_2 \xrightarrow[Et_3N]{t\text{-}BuPh_2SiCl} RNHSiPh_2t\text{-}Bu$$

These derivatives are stable to chromatography and to a variety of non-acidic reaction conditions, such as solvolysis (methanol or water/THF, 4:1, 6 h), and to aqueous base (20% KOH/MeOH, 16 h, reflux). They are also stable to LDA, to n-BuLi, and to NaBH₄/EtOH, and to reactive alkylating and acylating agents. Although stable to the conditions of Swern oxidation (DMSO, (COCl)₂, triethylamine), they do not survive other common oxidizing procedures. The primary amines are readily regenerated under mildly acidic conditions (AcOH/water, 4:1, 30 min, 25 °C, or THF/0.5 M HCl, 2:1, 6 h, 25 °C), or with pyridine–HF (1 eq., THF/water, 2 h, 25 °C).

General procedure (2)

A solution of the primary amine (10 mmol), t-butylchlorodiphenylsilane (10 mmol), and triethylamine (15 mmol) in MeCN (30 ml) was stirred at ambient temperature for 1–3 h. The reaction mixture was concentrated *in vacuo*, and the residue was partitioned between hexane/AcOEt (4:1), and 1 M sodium hydrogen carbonate solution. The organic phase was dried over a mixture of potassium carbonate and sodium sulphate.

Note. Secondary amines do not react under these conditions.

Secondary amines do not give the expected *N*-TBDMS derivatives under the conditions normally used for similar alcohol protection (Chapter 14). Instead, *N*-formamides are produced (3) in good yield through a DMF-derived Vilsmeier intermediate.

13.1.3. Alkylation of Ethyl Glycinate

N,N-Bis(trimethylsilyl) alkyl glycinates can be deprotonated and then alkylated on carbon (4) to give homologous α-amino-acid derivatives:

$$\underset{\overset{|}{NH_2}}{CH_2COOEt} \xrightarrow{Me_3SiNEt_2} \underset{\overset{|}{N(SiMe_3)_2}}{CH_2COOEt} \xrightarrow[2. RX]{1. NaN(SiMe_3)_2} \underset{\overset{|}{N(SiMe_3)_2}}{RCHCOOEt}$$

Ethyl glycinate was heated with excess TMSNEt₂ in the presence of a catalytic amount of ammonium sulphate at 120 °C, until no more diethylamine distilled out. The product was distilled directly (75%), b.p. 100 °C/10 mmHg.

To a solution of the silylated glycinate (50 mmol) in ether (50 ml), cooled to −10 to 0 °C, was added a solution of sodium hexamethyldisilazide (55 mmol) in ether (100 ml) with stirring. Stirring was continued at ambient temperature for a short time, and then the alkyl halide (50 mmol) was added dropwise. The mixture was heated under reflux for 10–15 h, cooled, filtered, and the product was distilled directly (52–70%).

Note. The alkyl halides used were MeI, EtI and PhCH₂Br.

REFERENCES

1. S. Djuric, J. Venit and P. Magnus, *Tetrahedron Lett.* **22**, 1787 (1981).
2. L. E. Overman, M. E. Okazaki and P. Mishra, *Tetrahedron Lett.* **27**, 4391 (1986).
3. S. W. Djuric, *J. Org. Chem.* **49**, 1311 (1984).
4. K. Rühlmann and G. Kuhrt, *Angew. Chem. Int. Edn* **7**, 809 (1968).

– 14 –

Alkyl Silyl Ethers

This chapter deals with the preparation, stability range, and cleavage of alkyl silyl ethers. Silyl ethers are used extensively in organic synthesis (1), and many interesting examples are contained in the preparative and cleavage references given. Many of the methods described are also applicable to other protic substrates, such as carboxylic acids, phenols and thiols, although individual stability ranges vary markedly. Preparations for most of the silylating agents cited here are detailed in Chapter 16.

14.1. PREPARATION

14.1.1. ROSiMe₃ (2)

$$2ROH \ + \ HMDS \ \xrightarrow{\text{TMSCl cat.}} \ 2ROSiMe_3 \ + \ NH_3 \uparrow$$

14.1.1.1. Primary and secondary

To a mixture of the alcohol (0.2 mol) and HMDS (0.11 mol) was added two drops of TMSCl, and the solution was gradually heated to reflux over 1–2 h. After an appropriate further time of reflux (*ca.* 4 h for primary and 16 h for secondary), the reaction mixture was cooled and distilled.

14.1.1.2. Tertiary

A solution of TMSCl (0.07 mol) in hexane (10 ml) was added dropwise with stirring to a mixture of the alcohol (0.2 mol) and HMDS (0.067 mol). The resulting mixture was heated under reflux for 5 h, cooled, filtered, and the precipitate was washed with pentane (2×50 ml). The organic extracts were concentrated and the residue was distilled.

14.1.1.3. Primary, secondary and tertiary with no catalyst (3)

This valuable method utilizes the O-TMS enol ethers derived from either pentane-2,4-dione or methyl acetoacetate, the former being the more reactive. Even t-alcohols are rapidly and quantitatively silylated in DMF at room temperature. A similar technique can be used to introduce the TBDMS group, although here ptsa catalysis is required (4).

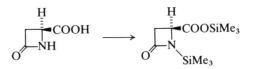

General procedure

4-Trimethylsilyloxypent-3-en-2-one (20.2 mmol) was added to a solution of the alcohol (20 mmol) in DMF (20 ml). After being shaken for 10 min, the mixture was extracted with pentane (5×10 ml). The combined organic extracts were washed with cold water (4×10 ml), dried and concentrated.

14.1.1.4. Trimethylsilyl esters

The use of HMDS (*ca.* 1.5 mmol) and saccharin (0.01 mmol) per mmol of substrate in refluxing dichloromethane or chloroform has been recommended (5) for easy silylation of carboxylic acids, including azetidin-2-one-4-carboxylic acids. Clear solutions result, i.e., no ammonium salts are present at completion of the reaction, and consequently the silyl esters can be obtained by direct distillation, or merely by evaporation of solvent.

Trimethylsilyl(4S)-N-(trimethylsilyl)azetidin-2-one-4-carboxylate (6)

To a suspension of (4*S*)-azetidin-2-one-4-carboxylic acid (0.1 mol) and saccharin (1 g) in chloroform (200 ml) was added HMDS (0.4 mol), and the mixture was heated under reflux for 1.5 h. The excess HMDS was removed under reduced pressure, and the residue was distilled to afford the protected β-lactam (0.089 mol, 89%), b.p. 74–76 °C/0.08 mmHg.

14.1.2. ROSiMe$_2$t-Bu

14.1.2.1. Primary and secondary

Procedure (i) (7)

A solution of the alcohol (1 mmol), TBDMSCl (1.2 mmol) and ImH (2.5 mmol) in DMF (2 ml/g alcohol) was stirred at room temperature for 24–48 h. The mixture was poured into pentane (25 ml), and washed with water, brine, and dried. Concentration followed by chromatography on SiO$_2$ or distillation gave the silyl ether.

Procedure (ii) (8)

A solution of the alcohol (1 mmol), TBDMSCl (1.1 mmol), Et$_3$N (1.2 mmol) and DMAP (0.04–0.1 mmol) in dichloromethane (5–10 ml) was stirred at ambient temperature for 24 h. The mixture was then diluted with dichloromethane (25 ml), washed with water and saturated ammonium chloride solution, and dried and concentrated.

14.1.2.2. Tertiary (9)

A solution of the alcohol (1 mmol), TBDMSOTf (1.5 mmol), and 2,6-lutidine (2 mmol) in dichloromethane (1 ml) was stirred for 10 min at ambient temperature. Normal extractive isolation followed by distillation gave the silyl ether (70–90%).

14.1.3. ROSi-i-Pr$_3$

14.1.3.1. Primary and secondary

Procedure (i) (9)

A solution of the alcohol (1 mmol), TIPSOTf (1.5 mmol) and 2,6-lutidine (2 mmol) in dichloromethane (1 ml) was stirred for 2 h at 0 °C. Normal extractive isolation and distillation gave the silyl ether.

Procedure (ii) (10)

A solution of the alcohol (1 mmol), TIPSCl (1.2 mmol) and ImH (2.5 mmol) in DMF (2 ml/g alcohol) was stirred at ambient temperature for 12–20 h. The mixture was poured into pentane (25 ml), and washed with water and brine, and dried. Concentration and distillation gave the product.

14.1.4. ROSiPh₂t-Bu (*11*)

14.1.4.1. Primary and secondary

A solution of the alcohol (1 mmol), TBDPSCl (1.2 mmol) and ImH (2.5 mmol) in DMF (2 ml/g alcohol) was stirred at room temperature for 24–48 h. The mixture was poured into pentane (25 ml), and washed with water and brine, and dried. Concentration followed by chromatography on SiO₂ or distillation gave the silyl ether.

14.1.5. ROSiMe₂t-Hex (*12*)

14.1.5.1. Primary and secondary

To a solution of thexyldimethylsilyl chloride (11 mmol) and ImH (15 mmol) in DMF (5 ml) was added the alcohol (11 mmol) at ambient temperature. After being stirred at ambient temperature for 16 h, the mixture was diluted with hexane. The hexane phase was washed with water (2×), and then dried. Concentration followed by distillation (Kugelrohr) gave the silyl ether (86–93%).

14.1.5.2. Tertiary

These undergo silylation using the more reactive silyl triflate in the presence of 2,6-lutidine or triethylamine as base in dichloromethane.

Notes

(*a*) TDSCl is cheaper to make on an industrial scale than TBDMSCl, since it avoids the use of t-BuLi.

(*b*) TDS ethers are approximately twice as stable as TBDMS ethers towards both acid- and fluoride-induced cleavage.

(*c*) Protection is afforded to amines, amides, thiols and carboxylic acids: ketones can be converted into the corresponding silyl enol ethers (triflate reagent).

(*d*) TDSOH, the hydrolysis product, can be recycled to TDSCl by heating with SOCl₂ or PCl₅ in 60–85% yield after distillation.

14.2. CLEAVAGE

Cleavage of trimethylsilyl ethers to the parent alcohols occurs quite readily on exposure to nucleophiles such as methanol, especially in the presence of acid or base. They do survive flash chromatography on silica gel. However, it is their general solvolytic lability that has led to the introduction of the more hindered, more stable silyl ethers detailed above.

14.2.1. Relative Rates of Cleavage

	$ArOSiR_3'$ $\longrightarrow$	$ArOH$
	Base-induced	Acid-induced
Me_3Si	32	1
Et_3Si	0.2	0.02
$t\text{-}BuMe_2Si$	1.6×10^{-3}	5.5×10^{-5}
	$ROSiR_3'$ $\longrightarrow$	ROH
Me_3Si	1	10^5
Et_3Si	10^{-3}	1.7×10^4
$i\text{-}Pr_3Si$		14

14.2.2. Stability Comparison between TBDMS, TIPS, and TBDPS Alkyl Ethers (10)

	Half-life of $R'OSiR_3$				
	$R' = n\text{-butyl}$		$R' = $ cyclohexyl		
R_3	H^+	HO^-	H^+	HO^-	F^-
TBDMS	<1 min	1 h	<4 min	26 h	76 min
TIPS	18 min	14 h	100 min	44 h	137 min
TBDPS	244 min	<4 h	360 min	14 h	not determined

H^+: 1% HCl/95% EtOH/22.5 °C
HO^-: 5% NaOH/95% EtOH/90 °C
F^-: 2 eq. n-Bu$_4$NF/THF/22.5 °C

Notes. (*a*) TIPS ethers are more stable in base than TBDMS ethers.
(*b*) TIPS ethers are hydrolysed more rapidly than TBDPS ethers in acid.

14.2.3. Cleavage of TBDMS Ethers (*13*)

General procedure

The TBDMS ether was dissolved in MeCN containing 5–30% of aqueous HF (40%), and the course of the reaction monitored by direct t.l.c. analysis. When deprotection was complete, chloroform and water were added. Normal isolation procedures then gave the free alcohol.

14.2.4. Selective Deprotection of Alcoholic and Phenolic TBDMS Ethers (*14*)

Under carefully controlled conditions, TBAF and aqueous HF selectively deprotect phenolic and alcoholic silyl ethers respectively. An excess of either reagent will, of course, ultimately result in complete deprotection.

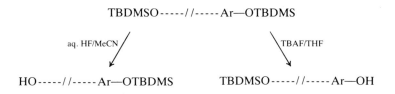

HF cleavage

An aqueous solution of HF (40%, 2 eq.) was added to a solution of the substrate in MeCN at either 0 °C or at ambient temperature. When the reaction was complete by t.l.c. analysis (*ca.* 30 min), excess sodium hydrogen carbonate solution (8%) was added, and the product was extracted with ether. The product was purified by flash chromatography on silica gel.

TBAF cleavage

A solution of TBAF in THF (1 eq., 1 M) was added to a solution of the substrate in THF at 0 °C. After *ca.* 15 min, an excess of saturated ammonium chloride solution was added, and the product was isolated as above.

14.2.5. Selective Deprotection of Bis(trimethylsilyl) Terminal Alkynols

Using sulphonic acid ion-exchange resins in ether solvent, selective removal of the trimethylsilyl group from oxygen in bistrimethylsilylated terminal alkynols can be achieved. This method is particularly suitable for low-molecular-weight compounds, where water solubility would make efficient extraction from an aqueous layer difficult.

6-Trimethylsilylhex-5-yn-1-ol (15)

$$Me_3SiC\equiv C(CH_2)_3CH_2OSiMe_3 \longrightarrow Me_3SiC\equiv C(CH_2)_3CH_2OH$$

To a vigorously stirred suspension of Rexyn 101 (Fisher Scientific, 7 ml, 14 meq. of H) in ether (10 ml) was added 1-trimethylsilyloxy-6-trimethyl-silylhex-5-yne (12.4 mmol). After 19 h, the reaction mixture was filtered through a glass frit, dried and concentrated. Distillation yielded 6-trimethyl-silylhex-5-yn-1-ol (11.7 mmol, 94%), b.p. 55–56 °C/0.5 mmHg.

Note. Substrate variation resulted in reaction times of 3–80 h, with reaction completion being monitored by g.l.c.

REFERENCES

1. *Reviews:* M. Lalonde and T. H. Chan, Use of organosilicon reagents as protective groups in organic synthesis, *Synthesis* 817 (1985); A. E. Pierce, *Silylation of Organic Compounds.* Pierce Chemical Co., Rockford, Illinois (1968); J. F. Klebe, *Adv. Org. Chem.* **8**, 97 (1972); B. E. Cooper, *Chem. Ind. (Lond.)* 794 (1978); T. W. Greene, *Protective Groups in Organic Synthesis.* Wiley, New York (1981).
2. S. H. Langer, S. Connell and I. Wender, *J. Org. Chem.* **23**, 50 (1958).
3. T. Veysoglu and L. A. Mitscher, *Tetrahedron Lett.* **22**, 1303 (1981).
4. T. Veysoglu and L. A. Mitscher, *Tetrahedron Lett.* **22**, 1299 (1981).
5. C. A. Bruynes and T. K. Jurriens, *J. Org. Chem.* **47**, 3966 (1982).
6. H. Fritz, P. Sutter and C. D. Weis, *J. Org. Chem.* **51**, 558 (1986).
7. E. J. Corey and A. Venkateswarlu, *J. Am. Chem. Soc.* **94**, 6190 (1972).
8. S. K. Chaudhary and O. Hernandez, *Tetrahedron Lett.* 99 (1979).
9. E. J. Corey, H. Cho, Ch. Rücker and D. H. Hua, *Tetrahedron Lett.* **22**, 3455 (1981).
10. R. F. Cunico and L. Bedell, *J. Org. Chem.* **45**, 4797 (1980).
11. S. Hanessian, *Acc. Chem. Res.* **12**, 159 (1979); S. Hanessian and P. Lavalee, *Can. J. Chem.* **53**, 2975 (1975).
12. H. Wetter and K. Oertle, *Tetrahedron Lett.* **26**, 5515 (1985); K. Oertle and H. Wetter, *Tetrahedron Lett.* **26**, 5511 (1985).
13. R. F. Newton, D. P. Reynolds, M. A. W. Finch, D. R. Kelly and S. M. Roberts, *Tetrahedron Lett.* 3981 (1979).
14. E. W. Collington, H. Finch and I. J. Smith, *Tetrahedron Lett.* **26**, 681 (1985).
15. R. A. Bunce and D. V. Hertzler, *J. Org. Chem.* **51**, 3451 (1986).

– 15 –

Silyl Enol Ethers and Ketene Acetals

15.1 PREPARATION

15.1.1. By Enolate Trapping

15.1.1.1. Silyl enol ethers

Various improvements in the regioselective synthesis of silyl enol ethers (*1*) have been described. The most frequently used route to silyl enol ethers is the trapping of enolate anions, generated under conditions of either kinetic or thermodynamic (equilibrium) control. The ability to obtain only one of the two regioisomeric silyl enol ethers derivable from an unsymmetrically substituted ketone is of critical importance to subsequent synthetic utility, and accordingly much effort has been expended in this direction. Yields of product are higher if a non-aqueous work-up procedure is used in the isolation of the less-substituted, kinetic silyl enol ether from the *O*-silylation of lithium enolates of unsymmetrical ketones, and prolonged heating increases the proportion of the more-substituted, thermodynamic silyl enol ether using triethylamine and TMSCl. This is exemplified (*2*) by the following reactions of 2-methylcyclohexanone; the regioisomeric ratios of the two silyl enol ethers are readily determined by [1]H n.m.r. spectroscopy:

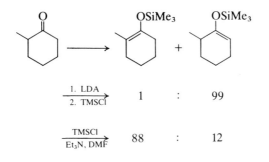

2-Methyl-1-(trimethylsilyloxy)cyclohex-6-ene (2)

A solution of 2-methylcyclohexanone (100 mmol) in THF (10 ml) was added to a stirred solution of LDA (from n-BuLi (106 mmol, 1.6 M in hexane) and diisopropylamine (120 mmol)) in THF (240 ml) over 10 min at −78 °C. The solution was stirred at −78 °C for a further 1 h, and then TMSCl (170 mmol) was added over 5 min. The solution was allowed to warm to ambient temperature, and, after being stirred for 1 h, it was then concentrated *in vacuo*. Pentane (100 ml) was added, and the precipitated LiCl removed by filtration through Celite. Concentration *in vacuo*, followed if necessary by re-filtration, and distillation gave the kinetic silyl enol ether (97 mmol, 97%), as essentially a single regioisomer, b.p. 77–78 °C/19 mmHg.

2-Methyl-1-(trimethylsilyloxy)cyclohex-1-ene (2)

TMSCl (150 mmol) was added dropwise with stirring to a solution of 2-methylcyclohexanone (125 mmol) and triethylamine (300 mmol) in DMF (50 ml). The mixture was heated in an oil bath at 130 °C for 90 h. On cooling, it was diluted with ether (200 ml), and washed with cold saturated sodium hydrogen carbonate solution (200 ml). The aqueous phase was re-extracted with ether (3 × 200 ml), and the combined organic extracts were washed rapidly and successively with dilute aqueous HCl (250 ml, 0.5 M), saturated sodium hydrogen carbonate solution (2 × 200 ml) and water (200 ml). After drying and concentration, distillation of the residue gave the thermodynamic silyl enol ether (104 mmol, 83%), in an enriched 88 : 12 regioisomeric ratio, b.p. 82–84 °C/16 mmHg.

A further improvement utilizes the compatibility of hindered lithium dialkylamides with TMSCl at −78 °C. Deprotonation of ketones and esters with lithium dialkylamides in the presence of TMSCl leads to enhanced selectivity (3) for the kinetically generated enolate. Lithium t-octyl-t-butylamide (4) appears to be superior to LDA for the regioselective generation of enolates and in the stereoselective formation of (E) enolates.

General in situ trapping procedure (3)

To a solution of the lithium dialkylamide (1.1 mmol) in THF (2 ml) cooled to −78 °C was added a solution to TMSCl (5–10 mmol) in THF (2 ml), also cooled to −78 °C. This was followed by dropwise addition of the carbonyl compound (1 mmol) in THF (2 ml). After 1 min, triethylamine (2 ml) was added, followed by quenching with saturated sodium hydrogen carbonate solution. The product was extracted into pentane, and these extracts were

washed successively with water and an aqueous solution of citric acid (0.1 M).
Drying and concentration gave the silyl enol ether.

15.1.1.2. Ketene silyl acetals

Two general methods can be used, the choice depending on whether the
parent ester is a dialkyl- or a monoalkylacetate. Many functional variations
can be tolerated, including monosubstituted malonates, and γ- and
δ-lactones (5).

(i) *Preparation of acetals $CH_2\!=\!C(OR)OSiR'_3$*

Acetates themselves normally give a mixture of *O*- and *C*-silylated products. However, using TBDMSCl in the presence of HMPA (CAUTION
—CANCER SUSPECT AGENT), pure ketene acetals of the type
$CH_2\!=\!C(OR)OTBDMS$ can be obtained (6).

1-t-Butyldimethylsilyloxy-1-ethoxyethene (6)

$$CH_3COOEt \xrightarrow[\text{2. TBDMSCl/HMPA}]{\text{1. LDA}} CH_2\!=\!C(OEt)OTBDMS$$

n-BuLi (1.5 M in hexane, 21 mmol) was treated dropwise with diisopropylamine (21 mmol) with stirring at 0 °C. The gel-like mixture was held at 0 °C
for 30 min, and then the hexane solvent was removed under reduced
pressure. The residual LDA was dissolved in THF (33 ml), and then cooled
to −78 °C. Ethyl acetate (20 mmol) was added dropwise, and stirring was
continued for 30 min. At this point, HMPA (CAUTION—CANCER
SUSPECT AGENT) (3.3 ml) was added dropwise, followed by a solution of
TBDMSCl (21 mmol) in pentane (10 ml). The stirred mixture was allowed to
come to ambient temperature overnight. The solvent was removed on a
rotary evaporator, and the residue was partitioned between pentane
(200 ml) and saturated sodium hydrogen carbonate solution (30 ml). The
pentane extract was washed with water and brine, and dried. Concentration
and distillation (Kugelrohr) gave the t-butyldimethylsilyl ketene acetal of
ethyl acetate (15.1 mmol, 76%), b.p. 175–180 °C/30 mmHg.

On the other hand, it has recently been reported (7) that phenyl acetate
forms the derived ketene trimethylsilyl acetal by exclusive *O*-silylation.

1-Phenoxy-1-trimethylsilyloxyethene (7)

$$CH_3COOPh \xrightarrow[\text{TMSCl}]{\text{LDA}} CH_2\!=\!C(OPh)OTMS$$

A mixture of phenyl acetate (50 mmol) and TMSCl (120 mmol) was added over 30 min to a pre-cooled (-80 °C) solution of LDA (75 mmol) in THF (40 ml). Stirring was continued at -80 °C for 30 min, and then the cooling bath was removed and the reaction mixture allowed to come to ambient temperature. It was then concentrated *in vacuo*, extracted with pentane, the pentane extract filtered and concentrated, and the residue distilled, to give the ketene acetal (37.5 mmol, 75%), b.p. 50–52 °C/0.1 mmHg.

(ii) *Preparation of acetals RR'C═C(OMe)OTMS (8)*

General procedure (8)

To a solution of LDA (from diisopropylamine (0.1 mol) and n-BuLi (0.1 mol)) in THF (75 ml) was added the ester $RR'CHCO_2Me$ (0.1 mol) at 0 °C over 5 min with stirring, which was continued for a further 30 min. Excess TMSCl (0.25 mol) was added over 5 min, and the reaction mixture was allowed to come to ambient temperature. After stirring for 30 min, the mixture was filtered using suction through a Celite pad, and the filtrate was concentrated using a rotary evaporator. The residue was taken up in ether, and filtration and concentration were repeated. Distillation of the residue under reduced pressure gave the ketene acetal, *ca.* 95%.

(iii) *Preparation of acetals RCH═C(OMe)OTMS (8)*

General procedure (8)

To a solution of LDA in THF prepared as above and cooled to -78 °C was added the ester RCH_2CO_2Me over 5 min with stirring, which was continued at -78 °C for a further 30 min. Excess TMSCl (0.25 mol) was added over 5 min, and the reaction mixture was allowed to come to ambient temperature. Isolation as above followed by distillation gave the product ketene acetal, *ca.* 90%.

(iv) *Ketene bis(trimethylsilyl)acetals, RR'C═C(OTMS)₂ (9)*

These can be prepared in good yield by reaction of either the mono-anions of trimethylsilyl carboxylates or, preferably, the dianions of carboxylic acids with TMSCl. Acetic and propionic acids give mixtures of *O*- and *C*-silylated products.

General procedure (9)

Sodium hydride (0.12 mol, 52% dispersion in mineral oil) was washed twice with ether to remove the oil, the ether being removed by decantation. Ether (200 ml) was added, and to the resulting suspension was added, with stirring, the carboxylic acid (0.1 mol) in ether (50 ml), at such a rate as to maintain gentle reflux. TMSCl (0.15 mol) was added dropwise, and the mixture was heated under reflux for 1–2 h. On cooling, it was filtered by suction through a pad of Celite, and the filtrate was carefully concentrated and the residue distilled, to give the trimethylsilyl carboxylate, *ca.* 80%.

A solution of LDA (0.1 mol) in THF (75 ml), prepared as above, was then cooled to −78 °C, and a solution of the TMS carboxylate (0.1 mol) in THF (40 ml) was added with stirring, which was continued at −78 °C for a further 30 min. Excess TMSCl (0.5 mol) was added over 5 min, and the reaction mixture was allowed to come to ambient temperature over 30 min with stirring. It was then filtered by suction through a pad of Celite, and concentrated using a rotary evaporator. The residue was taken up in ether (50 ml), and filtration and concentration were repeated. Distillation of the residue gave the ketene bis(trimethylsilyl)acetals, *ca.* 90%.

Alternatively, a solution of LDA (0.1 mol) in THF (75 ml), prepared as above, was cooled to 0 °C, and a solution of the carboxylic acid (0.05 mol) in THF (25 ml) was added dropwise with stirring, which was continued at 0 °C for a further 30 min. Excess TMSCl (0.25 mol) was added dropwise over 5 min, and the reaction mixture was allowed to come to ambient temperature over 30 min with stirring. Isolation as above, followed by distillation, gave the ketene bis(trimethylsilyl) acetals, *ca.* 90–95%.

15.1.2. By Hydrosilylation

Rhodium-catalysed addition (*10*) of hydridosilanes (Chapter 17) to αβ-unsaturated carbonyl compounds can be performed regioselectively, to afford either the product of 1,2-addition, or, perhaps more usefully, that of 1,4-addition, i.e. the corresponding silyl enol ether; this latter process is an excellent method for the regiospecific generation of silyl enol ethers. Of all catalyst systems investigated, tris(triphenylphosphine)rhodium(I) chloride proved to be the best.

15.1.2.1. Silyl enol ethers

General procedure (10)

A neat mixture of the $\alpha\beta$-unsaturated ketone (10 mmol), triethylsilane (11 mmol), and tris(triphenylphosphine)rhodium(I) chloride (0.01 mmol) was stirred at 50 °C for 2 h, and the product silyl enol ether was distilled directly (yields 90–98%).

Example of use with dimethylphenylsilane (11). A mixture of the $\alpha\beta$-unsaturated ketone (1.05 mmol), dimethylphenylsilane (1.1 mmol) and tris(triphenylphosphine)rhodium(I) chloride (0.002 mmol) was heated at 55 °C for 1 h. The silyl enol ether was distilled directly from the reaction.

15.1.2.2. Ketene silyl acetals

This reaction can also be used for the preparation of ketene silyl acetals from $\alpha\beta$-unsaturated esters, including simple acrylates (*12*); in geometrically defined cases, the (*Z*) isomer is produced stereoselectively (*7*):

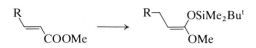

General procedure (12)

A mixture of the $\alpha\beta$-unsaturated ester (14 mmol), t-butyldimethylsilane (18 mmol) and tris(triphenylphosphine)rhodium(I) chloride (0.56 mmol) was placed in a pre-heated (100 °C) oil bath, and the course of reaction monitored by i.r. spectroscopy. On completion (*ca.* 30 min) the product was isolated by direct distillation (60–88%).

15.1.3. Asymmetric Synthesis of Trimethylsilyl Enol Ethers

Enantioselective deprotonation of prochiral 4-alkylcyclohexanones using certain lithium amide bases derived from chiral amines such as (**1**) has been shown (*13*) to generate chiral lithium enolates, which can be trapped and used further as the corresponding trimethylsilyl enol ethers; trapping was achieved using Corey's internal quench described above.

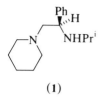

(1)

(R)-4-t-Butyl-1-(trimethylsilyloxy)cyclohexene (13)

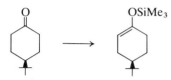

n-BuLi (1.5 M in hexane, 2.4 mmol) was added with stirring to a solution of the amine (1) (2.5 mmol) in THF (50 ml) at −78 °C. After 5 min, HMPA (CAUTION—CANCER SUSPECT AGENT) (4.8 mmol—sufficient to satisfy the tetracoordinate character of lithium) was added, the mixture was allowed to come to ambient temperature and then recooled to −78 °C. TMSCl (10 mmol) was added, followed by a solution of 4-t-butylcyclohexanone (2 mmol) in THF (4 ml) over 3 min. Stirring was continued at −78 °C for 10 min and then the mixture was quenched by the addition of triethylamine (4 ml) and saturated sodium hydrogen carbonate solution (10 ml). On reaching ambient temperature, the mixture was partitioned between pentane and saturated sodium hydrogen carbonate solution, and the crude product was isolated as above. Column chromatography followed by distillation (Kugelrohr) gave (R)-4-t-butyl-1-(trimethylsilyloxy)cyclohexene (1.34 mmol, 67%, 84% ee).

15.2. REACTIONS

The reactivity pattern (*1*) of silyl enol ethers and ketene acetals is based largely on their synthetic equivalence to enolate anions. Recently, a different spectrum of behaviour has been revealed, particularly in those reactions that involve direct reaction without prior generation of the enolate anion. Indeed, the historic development of silyl enol ethers can be seen in three separate phases, involving

(i) silylation as a trap for specific enolate ions, with subsequent enolate generation and reaction with electrophiles under *basic* conditions;

(ii) direct reaction of silyl enol ethers with electrophiles that are either Lewis *acids*, or can be made so by the addition of Lewis acid catalysts, and the use of fluoride ion to activate the silyl enol ether towards nucleophilic attack;

(iii) the use of silyl enol ethers in reactions to give products unobtainable by either of the two earlier phases.

15.2.1. Enolate Generation using MeLi

2-Allyl-2-methylcyclohexanone (14)

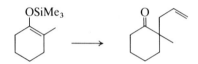

A solution of methyllithium–lithium bromide complex (30 mmol, 1.05 M in ether) was added with stirring at ambient temperature to 1-trimethyl-silyloxy-2-methylcyclohex-1-ene. After a further 30 min stirring, the reaction mixture was concentrated under reduced pressure on a rotary evaporator, the vacuum being released with nitrogen. THF (20 ml) was added, and then allyl bromide (29.7 mmol) was added dropwise with stirring at ambient temperature. After a further 5 min, the mixture was partitioned between pentane (100 ml) and saturated sodium hydrogen carbonate solution (50 ml). The separated organic layer was dried and concentrated under reduced pressure. Distillation gave the allyl ketone (16.5 mmol, 60%), b.p. 125–130 °C/25 mmHg.

15.2.2. Formal Enolate Generation by Fluoride Ion, Increasing the Nucleophilicity of the Silyl Enol Ether

Kuwajima (*15*) has provided full details of the regiospecific monoalkylation of carbonyl compounds *via* their silyl enol ethers, using stoichiometric amounts of fluoride ion. Noyori (*16*) has given more information on the use of the complex fluoride source (**2**) (Chapter 18)

$$(Et_2N)_3S^+ \quad Me_3SiF_2^-$$

(**2**)

to generate "naked" enolate anions. These react with a high degree of *syn* diastereoselectivity with aldehydes, regardless of the original enolate geometry, possibly through an extended acyclic transition state. The same fluoride source induces 1,4-addition of silyl ketene acetals to enones (*17*); a similar conjugate addition of ester enolates to acrylate esters has been explored (*18*) as a method of group-transfer polymerization.

It would appear that when fluoride ion is to be used in stoichiometric amounts, benzyltrimethylammonium fluoride is the preferred source; on the other hand, tetra-n-butylammonium fluoride, commercially available as its trihydrate, is more convenient in catalytic situations. However, there are difficulties (*19*) in successfully dehydrating the latter source without inducing Hofmann elimination.

Benzyltrimethylammonium fluoride (15)

A solution of benzyltrimethylammonium hydroxide (Triton B, 10 ml, 40% in MeOH) was treated with aqueous HF (*ca.* 8.6 ml, 4.7%) until the pH reached 8–7. The solvent was removed *in vacuo* (*ca.* 1 mm), and the residue was dried at 50 °C/0.5 mmHg for 20 h. The resulting highly hygroscopic solid was powdered, and then stored in a desiccator over P_2O_5.

2-(Methoxycarbonyl)methylcyclohexanone (15)

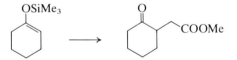

A mixture of finely ground benzyltrimethylammonium fluoride (125 mmol) and molecular sieves (50 g, 4 Å) in THF (50 ml) was stirred for 14 h at ambient temperature, and then the mixture was cooled to 0 °C. A solution of 1-(trimethylsilyloxy)cyclohexene (115 mmol) and methyl bromoacetate (105 mmol) in THF (50 ml) was added with stirring over 10 min. After stirring for a further 10 min at 0 °C, stirring was continued for 11 h at ambient temperature. The reaction mixture was filtered through Celite, concentrated and distilled, to give the ketoester (75 mmol, 65%), b.p. 90–100 °C/1.2 mmHg.

15.2.3. Lewis-Acid-Induced Reactions, Increasing the Electrophilicity of the Electrophile

15.2.3.1. TiCl₄-induced reactions with aldehydes and ketones

The Lewis acid encountered most often in this context is $TiCl_4$; its broad applicability in directed aldol reactions has been reviewed by Mukaiyama (20).

2-(Hydroxy)phenylmethylcyclohexanone (21)

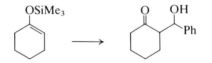

A solution of 1-(trimethylsilyloxy)cyclohexene (2.5 mmol) in dichloromethane (10 ml) was added dropwise to a solution of PhCHO (2.75 mmol) and $TiCl_4$ (2.75 mmol) in dichloromethane (20 ml), cooled to −78 °C, and the resulting mixture was stirred for 1 h. It was then quenched at this temperature by the addition of excess aqueous potassium carbonate solution, and allowed to come to ambient temperature. The mixture was extracted thoroughly with ether, and the combined organic extracts were washed with water and dried. After concentration under reduced pressure, the residue was purified by chromatography on silica gel, eluting with dichloromethane. This gave *erythro*-2-(hydroxy)phenylmethylcyclohexanone (0.56 mmol, 23%), m.p. 103.5–104.5 °C, and *threo*-2-(hydroxy)phenylmethylcyclohexanone (1.7 mmol, 68%), m.p. 75 °C.

15.2.3.2. TiCl₄-induced reactions with acetals and ketals

3,4-Diphenyl-4-ethoxybutan-2-one (22)

To a solution of $TiCl_4$ (2.6 mmol) in dichloromethane (5 ml), cooled to −78 °C, was added a solution of benzaldehyde diethyl acetal (2.5 mmol) in dichloromethane (10 ml). Immediately afterwards, a solution of 1-phenyl-2-trimethylsilyloxyprop-1-ene (2.5 mmol) in dichloromethane (10 ml) was added, and the mixture was stirred for 3 h at −78 °C. It was then quenched at this temperature by the addition of excess aqueous potassium carbonate solution, and allowed to come to ambient temperature. The mixture was extracted thoroughly with ether, and the combined organic extracts were washed with water and dried. After concentration under reduced pressure, the residue was purified by chromatography on silica gel, to give 3,4-diphenyl-4-ethoxybutan-2-one (2.375 mmol, 95%), as a mixture of diastereoisomers.

Notes. (*a*) Using trimethyl orthoformate, β-ketoacetals are obtained.

(*b*) The advantage of using acetals or ketals instead of aldehydes or ketones is that they react only as electrophiles, and probably coordinate more effectively with Lewis acids than do the parent carbonyl compounds.

15.2.3.3. Acylation

This succeeds well with ketone- and simple ester-derived enol ethers and ketene acetals (*23*).

2-Acetyl-2-methylcyclohexanone (23)

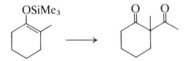

A solution of 2-methyl-1-(trimethylsilyloxy)cyclohex-1-ene (10 mmol) in dichloromethane (8 ml) was added dropwise over 15 min with stirring to acetyl chloride (10 mmol) and TiCl$_4$ (10 mmol) in dichloromethane (15 ml) at −78 °C. After 1 h at −78 °C, the reaction mixture was allowed to warm to ambient temperature over 2 h. It was then diluted with ether (20 ml), and poured into saturated sodium hydrogen carbonate solution (50 ml). Normal work-up, followed by chromatography on silica gel, gave the β-diketone (9.1 mmol, 91%).

Note. With silyl ketene acetals, anhydrous zinc bromide (1 mol%) is the preferred catalyst.

15.2.3.4. Conjugate addition to nitro-olefins—1,4-diketone synthesis

2-Acetonylcyclohexanone (24)

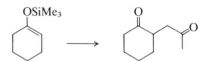

1-Trimethylsilyloxycyclohex-1-ene (1 mmol) was added dropwise to a solution of 2-nitropropene (1.5 mmol) and SnCl$_4$ (1 mmol) in dichloromethane (10 ml) at −78 °C over 5 min. After being stirred at −78 °C for 1 h, the mixture was allowed to warm to 0 °C over 2 h. Water (1.5 ml) was added, and the resulting mixture was stirred and heated under reflux for 2 h. On cooling, it was extracted with ethyl acetate (25 ml), and the organic extract was washed with water and brine. Removal of the solvent gave a residue, which was filtered through a short column of alumina (Woelm, activity III), eluting with ether. Distillation of the eluate gave pure 2-acetonylcyclohexanone (0.85 mmol, 85%), b.p. 80 °C/0.2 mmHg.

15.2.3.5. BF$_3$.OEt$_2$-induced reaction of acylsilane silyl enol ethers with acetals

This provides a route to αβ-unsaturated aldehydes (*25*).

2-Methyl-3-phenylprop-2-enal (25)

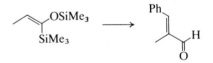

A solution of the trimethylsilyl enol ether of propionyl trimethylsilane (5 mmol) (Chapter 12) and benzaldehyde diethyl acetal (5 mmol) in dichloromethane (10 ml) was added to a solution of BF$_3$.OEt$_2$ (5 mmol) in dichloromethane (5 ml), cooled to −78 °C. After being stirred for 1 h at −78 °C and 2 h at −30 °C, the mixture was quenched with excess saturated sodium hydrogen carbonate solution, and extracted with ether. Concentration and distillation gave the product β-ethoxy acylsilane, (4.6 mmol, 95%), b.p. 105–106 °C/2 mmHg. Treatment of this alkoxy acylsilane (4 mmol) with tetrabutylammonium hydroxide (0.8 mmol, 25% solution in methanol) in acetonitrile (15 ml) for 15 min at ambient temperature, followed by the addition of ether (100 ml), neutralization with dilute aqueous HCl and washing with saturated sodium hydrogen carbonate solution, followed by concentration and distillation, gave 2-methyl-3-phenylprop-2-enal (3.76 mmol, 94%).

Notes. (*a*) With acetals, the product unsaturated aldehydes are usually exclusively (*E*).
(*b*) This reaction fails with ketals.

15.2.3.6. t-Alkylation and related reactions

Lewis-acid-induced alkylation reactions, employing S_N1 electrophiles, has been advanced and reviewed by Reetz (*26*) and Fleming (*27*) and their collaborators.

15.2.4. TMSOTf-Induced Reactions

15.2.4.1. TMSOTf-catalysed reaction with dimethyl acetals

2-(Methoxyphenylmethyl)cyclohexanone (28)

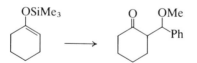

A solution of 1-trimethylsilyloxycyclohex-1-ene (5.12 mmol) and benzaldehyde dimethyl acetal (5.47 mmol) in dichloromethane (15 ml) was cooled to −78 °C, and to this. was added TMSOTf (0.05 mmol) in dichloromethane (0.5 ml). The mixture was stirred at −78 °C for 8 h, and then quenched by the addition of water at −78 °C. Dichloromethane (50 ml) was added, and the mixture was washed with saturated sodium hydrogen carbonate solution and brine, and dried. Concentration provided a crude oil consisting of a 93 : 7 mixture of *erythro-* and *threo*-2-(methoxyphenyl-methyl)cyclohexanone. Chromatography on silica gel (20 g, eluant petroleum ether : ether 10 : 1) gave the pure *erythro* (82%) and *threo* (6.7%) isomers as oils.

15.2.4.2. TMSOTf-catalysed reactions with 4-acetoxyazetidin-2-one

4-(Benzoylmethyl)azetidin-2-one (29)

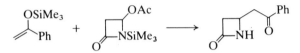

To a solution of 4-acetoxyazetidin-2-one (51.7 mmol) in ether (100 ml), cooled to 0 °C, was added triethylamine (61.2 mmol) followed by TMSCl (56 mmol). After stirring the reaction for 2 h at 0 °C, the solvent was removed *in vacuo* and the residue was extracted with pentane (4×60 ml). The combined extracts were filtered, concentrated and distilled to give *N*-trimethylsilyl-4-acetoxyazetidin-2-one (46.5 mmol, 90%), b.p. 80–81 °C/ 0.5 mmHg.

A solution of TMSOTf in dichloromethane (1% v/v, 1 ml) was added to a solution of the *β*-lactam (2.01 mmol) and 1-phenyl-1-trimethylsilyl-oxyethene (2.2 mmol) in dichloromethane, cooled to −78 °C. After being stirred at −78 °C for 15 min, the reaction mixture was allowed to warm to ambient temperature over 20 min, and was stirred for a further 30 min. The lime-green solution was quenched with aqueous KF (5% w/v, 20 ml) and extracted with dichloromethane (2×25 ml). Drying, concentration and chromatography of the residue gave the *β*-lactam (1.79 mmol, 89%) as a white solid, m.p. 141–143 °C.

15.2.5. Formal Cycloaddition Reactions

15.2.5.1. Diels–Alder reactions of silyloxydienes

The most frequently encountered, and most useful, cycloaddition reactions of silyl enol ethers are Diels–Alder reactions involving silyloxybutadienes (Chapter 18). Danishefsky (*30*) has reviewed his pioneering work in this area, and has extended his studies to include heterodienophiles, particularly aldehydes. Lewis acid catalysis is required in such cases, and substantial asymmetric induction can be achieved using either a chiral lanthanide catalyst or an *α*-chiral aldehyde.

15.2.6. [3,3]-Sigmatropic Rearrangement Reactions

15.2.6.1. Ireland–Claisen rearrangement of silyl ketene acetals (*31*)

Conversion of allyl alcohol esters into their corresponding trimethylsilyl or t-butyldimethylsilyl ketene acetals, followed by mild thermolysis, results in clean [3,3]-sigmatropic rearrangement to γδ-unsaturated silyl esters of carboxylic acids. Although trimethylsilyl ketene acetals rearrange at lower temperatures than do their t-butyldimethylsilyl analogues, involvement of the latter species is often preferable for two reasons. Not only is competitive α-C-silylation minimized, if not excluded, in their preparation, but also the intermediate silyl ketene acetals can be isolated and purified if desired.

In the course of this rearrangement, two new carbon–carbon bonds are formed, one single and one double. The new stereochemistries about both of these bonds are determined to a very large extent by the observation that, in the absence of unusual steric constraints, [3,3]-sigmatropic rearrangements proceed through chair-like transition states. As far as the double bond is concerned, this normally results in formation of the (E)-alkene isomer. A second consequence of a chair-like transition state is that the relative stereochemistry about the new single bond is predictable from the geometries of the double bonds of the allyl residue *and* of the ketene acetal.

(E)-Dec-4-enoic acid (31)

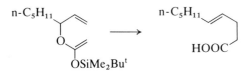

Hexane-free LDA was prepared by the dropwise addition of n-BuLi in hexane (1 eq, 2.4 M) to a stirred solution of di-isopropylamine (1.5 eq.) in hexane (*ca.* 2 M) at 0 °C. After complete addition, the viscous mixture was stirred for a further 10 min, after which time the hexane and excess di-iso-propylamine were removed under reduced pressure at 0 °C. The resulting white solid was redissolved in sufficient THF at 0 °C to give an approximately 0.3 M solution.

A solution of LDA (11 mmol) in THF (30 ml) was cooled to −78 °C, and HMPA (CAUTION—CANCER SUSPECT AGENT) (3 ml) then added. To this solution was added dropwise 3-acetoxyoct-1-ene (10 mmol), and then TBDMSCl (11 mmol) in THF (2 ml) over 5 min. The pale yellow solution was stirred at −78 °C for an additional 2 min, and the reaction mixture was allowed to warm to 25 °C over 30 min. It was stirred at this temperature for a further 2 h, and then quenched with water and pentane. The combined pentane extracts were concentrated, the crude oily silyl ester was dissolved in THF (25 ml) and dilute aqueous HCl (5 ml, 3 M) and the solution was then stirred for 45 min at 25 °C to complete hydrolysis. The mixture was then poured into aqueous sodium hydroxide (30 ml, 1 M) and extracted with ether (2×30 ml); the ether extracts were discarded. After acidification of the aqueous basic extract with concentrated HCl, the product acid was extracted with ether. After concentration, distillation of the residual oil gave the pure acid (8.3 mmol, 87%), b.p. 70 °C/ 0.003 mmHg.

Note. Less than 0.5% of the (*Z*) isomer was present (g.l.c. of methyl ester).

15.2.7. Oxidation

15.2.7.1. Using allyl carbonates

Silyl enol ethers and ketene acetals derived from ketones, aldehydes, esters and lactones are converted into the corresponding $\alpha\beta$-unsaturated derivatives on treatment with allyl carbonates in high yields in the catalytic presence of the palladium–bis(diphenylphosphino)ethane complex (*32*). A phosphine-free catalyst gives higher selectivity in certain cases, such as those involving ketene acetals. Nitrile solvents, such as acetonitrile, are essential for success.

$$\text{\Large$\diagdown$}\!\!-\!\text{OSiMe}_3 \quad \longrightarrow \quad \text{\Large$\diagdown$}\!\!=\!\text{O}$$

General procedure for silyl enol ethers (32)

A solution of Pd(OAc)$_2$ (0.05 mmol) and bis(diphenylphosphino)ethane (0.05 mmol) in acetonitrile (1 ml) was heated gently to reflux, at which time a solution of the silyl enol ether (1 mmol) and diallyl carbonate (2 mmol) in MeCN (4 ml) was added in one portion. The mixture was heated under reflux for 1–3 h, the course of reaction being monitored by t.l.c. or g.l.c. analysis. On completion, the cooled reaction solution was filtered through fluorosil. The pure $\alpha\beta$-unsaturated compound was isolated by column chromatography on silica gel (70–95%).

General procedure for ketene silyl acetals (32)

A solution of Pd(OAc)$_2$ (0.1 mmol), the ketene acetal (1 mmol) and allyl methyl carbonate (2 mmol) in MeCN (5 ml) was heated under reflux for 2–6 h, the course of reaction being monitored by t.l.c. or g.l.c. analysis. On completion, the cooled reaction solution was filtered through fluorosil. The pure αβ-unsaturated compound was isolated by column chromatography on silica gel (70–90%).

Note. Enol acetates undergo a similar oxidation, using the palladium species and tributyltin methoxide as dual catalysts.

15.2.7.2. Using lead(IV) acetate—Oxidation to α-acetoxyaldehydes (*33*)

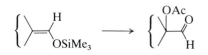

General procedure

A mixture of lead(IV) acetate (20 mmol) and KOAc (100 mmol) in AcOH (30 ml) was treated with the neat aldehyde-derived silyl enol ether (20 mmol) at ambient temperature. After being stirred for 1 h at ambient temperature, the reaction mixture was diluted with water (30 ml), and then extracted with pentane (3×200 ml). The combined pentane extracts were washed with saturated sodium hydrogen carbonate solution (2×50 ml), dried, concentrated and distilled to give the product α-acetoxyaldehyde (45–78%).

Note. Omission of KOAc results in extensive elimination by-products.

15.2.7.3. Using mcpba—Oxidation to α-trimethylsilyloxyketones (*34*)

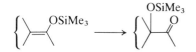

General procedure (34)

The trimethylsilyloxyalkene was treated with a slight excess of mcpba in dichloromethane at 0 °C. Addition of pentane, followed by filtration and rapid chromatography (neutral alumina, eluting with ether), gave the α-trimethylsilyloxyketones (60–75%).

15.2.8. Asymmetric Aldol

This area of reactivity has been the subject of excellent reviews (*35*). Silyl enol ethers are not sufficiently nucleophilic to react spontaneously with carbonyl compounds; they do so under the influence of either Lewis acids or fluoride ion, as detailed above. Few clear trends have emerged from the somewhat limited number of definitive studies reported so far, with ambiguities in diastereoisomeric assignments occasionally complicating the issue even further.

REFERENCES

1. *Reviews:* J. K. Rasmussen, *Synthesis* 91 (1977); I. Fleming, *Chimia* **34**, 265 (1980); P. Brownbridge, *Synthesis* 1 and 85 (1983); R. J. K. Taylor, Organocopper conjugate addition–enolate trapping reactions, *Synthesis* 364 (1985).
2. G. Stork and P. F. Hudrlik, *J. Am. Chem. Soc.* **90**, 4462, 4464 (1968); H. O. House, L. J. Czuba, M. Gall and H. D. Olmstead, *J. Org. Chem.* **39**, 2324 (1969): modified procedures by I. Fleming and I. Paterson, *Synthesis* 736 (1979).
3. E. J. Corey and A. W. Gross, *Tetrahedron Lett.* **25**, 495 (1984).
4. E. J. Corey and A. W. Gross, *Tetrahedron Lett.* **25**, 491 (1984).
5. J. K. Rasmussen and A. Hassner, *J. Org. Chem.* **39**, 2558 (1974); N. L. Holy and Y. F. Wang, *J. Am. Chem. Soc.* **99**, 944 (1977); I. Paterson and I. Fleming, *Tetrahedron Lett.* 993 (1979).
6. M. W. Rathke and D. F. Sullivan, *Synth. Commun.* **3**, 67 (1973): modified procedure by E. W. Colvin.
7. N. Slougi and G. Rousseau, *Synth. Commun.* **17**, 1 (1987).
8. C. Ainsworth and Y.-N. Kuo, *J. Organometal. Chem.* **46**, 73 (1972).
9. C. Ainsworth, F. Chen and Y.-N. Kuo, *J. Organometal. Chem.* **46**, 59 (1972).
10. I. Ojima and T. Kogure, *Organometallics* **1**, 1390 (1982).
11. G. S. Cockerill, P. Kocienski and R. Threadgold, *J. Chem. Soc. Perkin Trans. I* 2093 (1985).
12. E. Yoshii, Y. Kobayashi, T. Koizumi and T. Oribe, *Chem. Pharm. Bull.* **22**, 2767 (1974).
13. R. Shirai, M. Tanaka and K. Koga, *J. Am. Chem. Soc.* **108**, 543 (1986).
14. E. W. Colvin, unpublished.
15. I. Kuwajima, E. Nakamura, and M. Shimuzu, *J. Am. Chem. Soc.* **104**, 1025 (1982).
16. R. Noyori, I. Nishida and J. Sakata, *J. Am. Chem. Soc.* **105**, 1598 (1983).
17. T. V. RajanBabu, *J. Org. Chem.* **49**, 2083 (1984).
18. O. W. Webster, W. R. Kertler, D. Y. Sogah, W. B. Farnham and T. V. RajanBabu, *J. Am. Chem. Soc.* **105**, 5706 (1983).
19. R. K. Sharma and J. L. Fry, *J. Org. Chem.* **48**, 2112 (1983).
20. T. Mukaiyama, *Angew. Chem. Int. Edn* **16**, 817 (1977); see also Chapter 19 of the present book.
21. T. Mukaiyama, K. Banno and K. Narasaka, *J. Am. Chem. Soc.* **96**, 7503 (1974).
22. T. Mukaiyama and M. Hayashi, *Chem. Lett.* 15 (1974).
23. I. Fleming, J. Iqbal and E.-P. Krebs, *Tetrahedron* **39**, 841 (1983); for the use of carboxylic acid anhydrides with TMSOTf as catalyst, see R. Noyori, S. Murata and M. Suzuki, *Tetrahedron* **37**, 3899 (1981).

24. M. Miyashita, T. Yanami, T. Kumazawa and A. Yoshikoshi, *J. Am. Chem. Soc.* **106**, 2149 (1984); see also Chapter 18 of the present book.
25. T. Sato, M. Arai and I. Kuwajima, *J. Am. Chem. Soc.* **99**, 5827 (1977).
26. M. T. Reetz, *Angew. Chem. Int. Edn* **21**, 96 (1982); see also Chapter 18 of the present book.
27. I. Fleming, *Chem. Soc. Rev.* **10**, 83 (1981).
28. S. Murata, M. Suzuki and R. Noyori, *J. Am. Chem. Soc.* **102**, 3248 (1980).
29. R. P. Attrill, A. G. M. Barrett, P. Quayle, J. van der Westhuizen and M. J. Betts, *J. Org. Chem.* **49**, 1679 (1984).
30. S. Danishefsky, *Acc. Chem. Res.* **14**, 400 (1981); see also S. Danishefsky and T. A. Craig, *Tetrahedron*, **37**, 4081 (1981); S. Danishefsky, M. Bednarski, T. Izawa and C. Maring, *J. Org. Chem.* **49**, 2290 (1984); S. J. Danishefsky, E. Larson, D. Askin and N. Kato, *J. Am. Chem. Soc.* **107**, 1246 (1985); S. Danishefsky and J. F. Kerwin, *J. Org. Chem.* **47**, 1597 (1982); S. Danishefsky, S. Kobayashi and J. F. Kerwin, *J. Org. Chem.* **47**, 1981 (1982); S. J. Danishefsky, W. H. Pearson, D. F. Harvey, C. J. Maring and J. P. Springer, *J. Am. Chem. Soc.* **107**, 1256 (1985); M. Bednarski, C. J. Maring and S. Danishefsky, *Tetrahedron Lett.* **24**, 3451 (1983); M. Bednarski and S. Danishefsky, *J. Am. Chem. Soc.* **105**, 6968 (1983); S. J. Danishefsky and C. J. Maring, *J. Am. Chem. Soc.* **107**, 1269 (1985); S. J. Danishefsky, E. Larson and J. P. Springer, *J. Am. Chem. Soc.* **107**, 1274 (1985).
31. R. E. Ireland, R. H. Mueller and A. K. Willard, *J. Am. Chem. Soc.* **98**, 2868 (1976); for a chiral version, see R. E. Ireland and M. D. Varney, *J. Am. Chem. Soc.* **106**, 3668 (1984).
32. I. Minami, K. Takahashi, I. Shimuzu, T. Kimura and J. Tsuji, *Tetrahedron* **42**, 2971 (1986).
33. G. Rubottom, R. Marrero and J. M. Gruber, *Tetrahedron* **39**, 861 (1983); see also Chapter 18 of the present book.
34. A. G. Brook and D. M. Macrae, *J. Organometal. Chem.* **77**, C19 (1974); see also Chapter 18 of the present book.
35. C. H. Heathcock, *Asymmetric Synthesis*, ed. J. D. Morrison, Vol. 3, pp. 111–212, Academic Press, London (1984); D. A. Evans, J. V. Nelson and T. R. Taber, *Top. Stereochem.* **13**, 1 (1982); see, however, I. R. Silverman, C. Edington, J. D. Elliott and W. S. Johnson, *J. Org. Chem.* **52**, 180 (1987).

– 16 –

Silyl-Based Reagents

Formal replacement of the proton of certain inorganic acids generates a group of reagents (*1*) that can act as strong electrophiles, particularly when Si—O bond formation takes place. In those cases where the anion of the parent acid is a good nucleophile, the anion can attack the cationic species so formed, resulting in overall conversions such as ester and ether cleavage, and addition to aldehydes and ketones. Most also act as good silylating agents for alcohols (Chapter 14). For examples of other uses of these reagents, see Chapter 18.

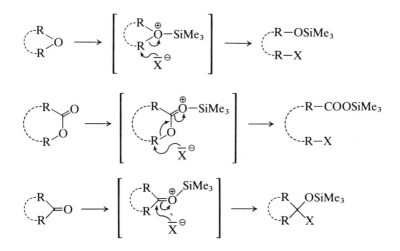

TMSOTf (2)

TfOH (10 mmol) and tetramethylsilane (12.5 mmol) were mixed at ambient temperature. After 1 h, evolution of methane had ceased. Direct distillation of the mixture gave the product (79% based on tetramethylsilane), b.p. 40 °C/11 mmHg, purity 96% (g.l.c.).

119

TBDMSOTf (3)

TfOH (0.16 mol) was added dropwise to TBDMSCl (0.16 mol) at 23 °C, and the resulting mixture was heated at 60 °C for *ca.* 10 h, by which time no further HCl was evolved. The product (0.128 mol, 80%) was distilled directly from the reaction flask, b.p. 60 °C/7 mmHg.

TIPSOTf (3)

TfOH (0.266 mol) was added dropwise to TIPSH (0.242 mol), with stirring and cooling at 0 °C. After completion of addition, the mixture was stirred for 16 h at ambient temperature and the product (0.234 mol, 97%) was distilled directly, b.p. 83–87 °C/1.7 mmHg.

TIPSH (4)

A solution of the Grignard reagent from i-PrCl (8 mol) and Mg turnings (8.3 g atom) in THF (4 l) was treated with Cl_3SiH (2 mol) at 0 °C, and the mixture was stirred for 3 days at 25 °C. Work-up and distillation gave the product (1.6 mol, 80% based on Cl_3SiH), b.p. 53–63 °C/18 mmHg.

TIPSCl (5)

TIPSH (0.25 mol) was added to a mixture of anhydrous copper(II) chloride (0.55 mol) and MeCN (250 ml), resulting in a two-phase liquid system. This was heated under reflux for 16 h, and then chilled to precipitate copper(I) chloride. The supernatant, consisting of a dark MeCN layer and a colourless phase of TIPSCl, was separated, and the MeCN phase was extracted with pentane. The pentane extract and the TIPSCl phase were combined, concentrated and distilled, to give the product (0.25 mol, 100%), b.p. 88–92 °C/18 mmHg.

TDSCl (6)

To a mixture of Me_2SiClH (92 mmol) and $AlCl_3$ (5 mmol) was added 2,3-dimethylbut-2-ene (91 mmol) at ambient temperature. The reaction mixture was stirred for 4 h at ambient temperature, and then filtered and distilled, to provide the product (84 mmol, 93%), b.p. 55–56 °C/10 mmHg.

TDSOTf (7)

Equimolar amounts of TDSCl and TfOH were heated at 60 °C for 5 h. The product was isolated by distillation, b.p. 40–42 °C/0.15 mmHg.

4-Trimethylsilyloxypent-3-en-2-one (8)

TMSCl (0.25 mol) was added to a stirred solution of ImH (0.223 mol) in THF (125 ml). A jelly-like precipitate formed. Pentane-2,4-dione (0.220 mol) was added *via* a hypodermic syringe, and then the septum was replaced with a glass stopper. The mixture was heated under gentle reflux with stirring overnight. On cooling, it was filtered through a sintered glass funnel filled with anhydrous sodium sulphate. Concentration of the filtrate followed by distillation gave the silyl enol ether (0.190 mol, 86%), b.p. 63 °C/3 mmHg.

TMSCN (9)

A mixture of dry KCN (100 mmol), TMSCl (110 mmol, distilled from calcium hydride), and 18-Crown-6 (0.4 mmol) in dichloromethane (20 ml) was heated under reflux for 24 h. Direct distillation using a 12.5 cm Vigreux column gave the product (36 mmol, 33%), b.p. 114–117 °C/760 mmHg.

TBDMSCN (10)

A mixture of dry KCN (1 mol), TBDMSCl (300 mmol), and 18-Crown-6 (7.6 mmol) in dichloromethane (200 ml) was heated under reflux with vigorous stirring for 48 h. The mixture was cooled and filtered, and the solvent was totally removed by distillation in a nitrogen atmosphere. Sublimation of the residue gave the product (267 mmol, 89%), m.p. 72–77 °C.

TMS diazomethane (11)

$$Me_3SiCH_2MgCl \ + \ (PhO)_2P(O)N_3 \ \longrightarrow \ Me_3SiCHN_2$$

The Grignard reagent prepared from chloromethyltrimethylsilane (30 mmol) and Mg (36 mg atom) in ether (20 ml) was added dropwise to diphenyl phosphorazidate (27 mmol) in ether (40 ml), keeping the temperature below 0 °C. The reaction mixture was stirred for 2 h at 0 °C, and then at ambient temperature for 3 h. It was then cooled to 0 °C, and ice-water was added. The mixture was filtered, and the solid was washed with ether. The combined ethereal extracts were washed with ice-water and dried. Careful concentration at <45 °C/atmospheric pressure, then distillation at

0–30 °C (bath temperature)/15 mmHg, gave a concentrated ethereal solution of the diazoalkane. To this was added hexane (10 ml), and the solution was concentrated at atmospheric pressure to give a solution of the diazoalkane in hexane (9.49 ml). ^{1}H n.m.r. indicated a concentration of 2.25 mmol/ml (79%).

2-Trimethylsilylethanol (12)

$$Me_3SiCH_2Cl \xrightarrow[\text{2. CH}_2\text{O}]{\text{1. Mg}} Me_3SiCH_2CH_2OH$$

A mixture of Mg (0.163 g atom) and chloromethyltrimethylsilane (0.163 mol) in ether (90 ml) was heated gently (air-gun) to initiate reaction (once started, the reaction can be very exothermic, and it should be cooled occasionally to maintain a gentle reflux). When all the Mg had dissolved, paraformaldehyde (5 g, 0.166 mol of CH_2O) was added, and reflux was continued for a further 2.25 h. The cooled mixture was diluted with ether, and washed with saturated ammonium chloride solution. The aqueous layer was re-extracted with ether, and the combined ethereal extracts were dried and concentrated. Distillation gave 2-trimethylsilylethanol (0.156 mol, 96%), b.p. 34–35 °C/1 mmHg.

Preparation of 2-trimethylsilylethyl esters (13)

$$RCOOH + Me_3SiCH_2CH_2OH \longrightarrow RCOOCH_2CH_2SiMe_3$$

To a solution of the carboxylic acid (2.1 mmol) in THF (40 ml) was added 2-trimethylsilylethanol (4.2 mmol) and TMSCl (16.5 mmol). The mixture was stirred under reflux for 36 h, cooled, and concentrated *in vacuo*. Direct chromatographic purification gave the ester (73–98%).

Note. This excellent method, which proceeds with the intermediacy of the trimethylsilyl carboxylates, is equally applicable to esterifications of acids with simple alcohols, which are then used as solvent.

Ketalization under aprotic conditions: benzaldehyde dimethyl acetal (14)

One of the most reliable reported methods for ketalization under anhydrous and non-equilibrating conditions is the TMSOTf-catalysed reaction with alkoxytrimethylsilanes *(14)*. Both aldehydes and ketones react smoothly with stoichiometric amounts of alkoxytrimethylsilanes in dichloromethane at temperatures as low as −78 °C.

$$\text{PhCHO} \xrightarrow[\text{TMSOTf cat.}]{\text{MeOSiMe}_3} \text{PhCH(OMe)}_2$$

To a stirred solution of TMSOTf (0.1 mmol) in dichloromethane (1 ml), cooled to −78 °C, were added successively methoxytrimethylsilane (20 mmol) and benzaldehyde (10 mmol). The mixture was stirred at −78 °C for 3 h, and then quenched by the addition of pyridine (0.2 ml) at −78°C. The mixture was poured on to saturated sodium hydrogen carbonate solution (15 ml), and extracted with ether (3 × 15 ml). The combined organic extracts were dried over sodium carbonate/sodium sulphate (1 : 1), filtered, concentrated and distilled, to give benzaldehyde dimethyl acetal (9.4 mmol, 94%), b.p. 125–135 °C/51 mmHg.

Notes. (*a*) 1,2-Bis(trimethylsilyloxy)ethane reacts similarly to give the corresponding dioxolanes.

(*b*) Notably, enones such as cyclohex-2- and cyclohex-3-enone undergo ketalization *without* concomitant double-bond migration.

Bis(trimethylsilyl) peroxide (15)

To a solution of DABCO (89 mmol) in THF (150 ml) was added dropwise, at 0 °C, hydrogen peroxide (23 ml, 27%, 180 mmol). The DABCO.2H$_2$O$_2$ complex, which precipitated immediately as a white solid, was filtered and dried *in vacuo*. A mixture of this complex (55.5 mmol) and DABCO (89 mmol) was then kept in a reaction flask for 2 h at 40 °C/1 mmHg to remove any traces of moisture. After cooling to 0 °C, dichloromethane (300 ml) was added, followed by TMSCl (220 mmol). The mixture was stirred for 3 h at 0 °C, and then filtered from the copious precipitate of DABCO.HCl. After concentration *in vacuo*, pentane (200 ml) was added, and the mixture was filtered once again. Careful concentration *in vacuo* gave the peroxide (96%, 98% pure by ^{1}H n.m.r.).

Ethyl trimethylsilylacetate (16)

$$\text{BrCH}_2\text{COOEt} + \text{TMSCl} \xrightarrow{\text{Zn}} \text{Me}_3\text{SiCH}_2\text{COOEt}$$

From a mixture of freshly sandpapered strips of Zn (0.5 g atom) in benzene (500 ml) 75 ml of the latter was distilled to ensure complete dryness. A solution of TMSCl (0.4 mol) and ethyl bromoacetate (0.5 mol) in benzene/ether (200 ml, 1 : 1) was added over 0.5 h to the gently refluxing, mechanically stirred mixture (during the addition, a crystal of I$_2$ was added to help initiate the reaction, which sometimes became vigorous, requiring cooling). After final addition, the reaction mixture was heated under reflux for 1–3 h, until all Zn had dissolved. It was then cooled in an ice bath, and dilute HCl (400 ml, 1 M) was added over 15 min with stirring. After a further 5 min stirring, the layers were separated. The organic layer was washed with dilute HCl (100 ml, 1 M), and the combined aqueous layers were extracted with ether. The combined organic layers were washed with water, saturated sodium hydrogen carbonate solution and water, and dried. Concentration and distillation gave ethyl trimethylsilylacetate (0.288 mol, 72%), b.p. 76–77 °C/40 mmHg.

Note. Omission of ether resulted in a lowered yield of 55%.

3-Trimethylsilylprop-2-yn-1-ol (17)

$$HC\equiv CCH_2OH \longrightarrow HC\equiv CCH_2OTHP \longrightarrow Me_3SiC\equiv CCH_2OH$$

A solution of propyn-1-ol (0.4 mol) and dihydropyran (1.1 mol) in dry dichloromethane (350 ml) containing pyridinium tosylate (2.5 g) was stirred at ambient temperature for 4 h. The solution was diluted with ether (400 ml), washed with saturated sodium hydrogen carbonate solution (2 × 100 ml) and brine (1 × 100 ml), and dried. Concentration and distillation gave the tetrahydropyranyl ether (0.312 mol, 78%), b.p. 74–76 °C/20 mmHg. To a solution of EtMgBr (from Mg (0.212 g atom) and EtBr (0.28 mol)) in ether (120 ml) was added the tetrahydropyranyl ether (0.2 mol) with vigorous stirring: the solid mass that developed was broken up with a spatula. Stirring was continued for 12 h at ambient temperature, and then TMSCl (0.33 mol) was added in approximately 10 ml portions. The resulting mixture was heated under gentle reflux for 12 h, by which time it had turned purple. Heating was discontinued, TMSCl (0.16 mol) was added, and stirring was maintained for a further 12 h at ambient temperature. Aqueous HCl (100 ml, 1 M) was added (*care*: exothermic), and stirring was continued for a further 2 h, by which time all solids had dissolved. The layers were separated, the aqueous layer was extracted with ether (3 × 100 ml), and the combined organic layers were washed with water (3 × 100 ml) and brine (3 × 100 ml), and dried. Concentration and distillation gave 3-trimethylsilylpropyn-1-ol (0.154 mol, 77%), b.p. 82–83 °C/20 mmHg.

REFERENCES

1. *Reviews*
   TMSCN, TMSI, TMSN₃ TMSSMe: W. C. Groutas and D. Felker, *Synthesis* 861 (1980).
   TMSI, TMSBr: A. H. Schmidt, *Aldrichimica Acta* **14**, 31 (1981).
   TMSI: G. A. Olah and S. C. Narang, *Tetrahedron* **38**, 2225 (1982).

TMSOTf: R. Noyori, S. Murata and M. Suzuki, *Tetrahedron* **37**, 3899 (1981); H. Emde, D. Domsch, H. Feger, U. Frick, A. Götz, H. H. Hergott, K. Hofmann, W. Kober, K. Krägeloh, T. Oesterle, W. Steppan, W. West and G. Simchen, *Synthesis* 1 (1982). *TMSCH₂Cl*: R. Anderson, *Synthesis* 717 (1985).

2. M. Demuth and G. Mikhail, *Synthesis* 827 (1982).
3. E. J. Corey, H. Cho, Ch. Rücker and D. H. Hua, *Tetrahedron Lett.* **22**, 3455 (1981).
4. R. F. Cunico and L. Bedell, *J. Org. Chem.* **45**, 4797 (1980).
5. R. F. Cunico and E. M. Dexheimer, *Synth. React. Inorg. Met.-Org. Chem.* **4**, 23 (1974).
6. K. Oertle and H. Wetter, *Tetrahedron Lett.* **26**, 5511 (1985).
7. H. Wetter and K. Oertle, *Tetrahedron Lett.* **26**, 5515 (1985).
8. T. Veysoglu and L. A. Mitscher, *Tetrahedron Lett.* **22**, 1303 (1981); for an alternative procedure, see D. T. W. Chu and S. N. Huckin, *Can. J. Chem.* **58**, 138 (1980).
9. J. W. Zubrick, B. I. Dunbar and H. D. Durst, *Tetrahedron Lett.* 71 (1975).
10. P. G. Gassman and L. M. Haberman, *J. Org. Chem.* **51**, 5010 (1986).
11. S. Mori, I. Sakai, T. Aoyama and T. Shioiri, *Chem. Pharm. Bull.* **30**, 3380 (1982).
12. M. L. Mancini and J. F. Honek, *Tetrahedron Lett.* **23**, 3249 (1982).
13. M. A. Brook and T. H. Chan, *Synthesis* 201 (1983); for a similar use of TMSCl in the facile preparation of ethylene ketals and other acetals, see T. H. Chan, M. A. Brook and T. Chaly, *Synthesis* 203 (1983).
14. T. Tsunoda, M. Suzuki and R. Noyori, *Tetrahedron Lett.* **21**, 1357 (1980).
15. M. Taddei and A. Ricci, *Synthesis* 633 (1986).
16. R. J. Fessenden and J. S. Fessenden, *J. Org. Chem.* **32**, 3535 (1967); see also Chapter 18 of the present book.
17. E. W. Colvin, unpublished; see also Chapter 18 of the present book for an alternative procedure.

– 17 –

Silanes as Reducing Agents

17.1. HYDROSILYLATION

The transition metal catalysed addition of a hydridosilane to a multiply-bonded system is known as hydrosilylation (*1*). Under such conditions, alkynes undergo clean *cis*-addition, so providing one of the most direct routes to vinylsilanes (Chapter 3). Hydridosilanes also add to the carbonyl group of saturated aldehydes and ketones, to produce alkyl silyl ethers. For example, under suitable conditions, 4-t-butylcyclohexanone (*2*) can be reduced with a high degree of stereoselectivity.

trans-(4-t-Butylcyclohexyloxy)triethylsilane (2)

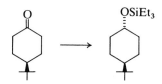

To a solution of 4-t-butylcyclohexanone (1 mmol), tris(triphenylphosphine)rhodium(II) chloride (0.05 mmol) and silver trifluoroacetate (0.05 mmol) in toluene (5 ml) was added triethylsilane (1.5 mmol). The mixture was heated under reflux for 20 h, and concentrated under reduced pressure. The residue was diluted with hexane (3 ml), filtered and distilled to give a mixture of triethylsilyl ethers (0.96 mmol, 96%), b.p. 70 °C/0.1 mmHg. G.l.c. analysis shows an axial (*cis*) : equatorial (*trans*) ratio of 5 : 95—a result comparable to the best LAH results.

Note. Use of tris(triphenylphosphine)rhodium(I) chloride/triethylsilane gave somewhat poorer results in terms of stereoselectivity, *ca.* 10 : 90.

Full details have been given (*3*) for the selective 1,2- or 1,4-reduction of αβ-unsaturated ketones. Dihydridosilanes give products of predominant

1,2-addition, whereas bulky monohydridosilanes favour products of 1,4-addition to give silyl enol ethers (Chapter 15). These observations have been rationalized in terms of competitive rates of hydrogen transfer from rhodium to carbon in the allyl intermediate (1) and of isomerization of (1) to (2); the latter isomerization will be accelerated by steric hindrance in (1) when bulky hydridosilanes are used:

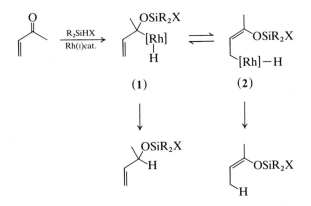

17.2. IONIC HYDROGENATION—DEOXYGENATION OF LACTOLS

As with the similarly polarized boron and aluminium hydrides, hydrido-silanes can transfer formal hydride ions to electropositive carbon centres. Unlike the first two reducing agents, hydridosilanes require additional activation of the carbon centre by Lewis or protic acids before such hydride transfer can take place. This overall process is known as ionic hydrogenation (4). The reagent system of triethylsilane and boron trifluoride etherate has provided an extremely selective method for the reductive deoxygenation of lactols (5), derived in turn from DIBAL reduction (6) of the corresponding γ- or δ-lactones:

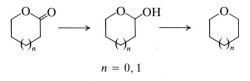

$$n = 0, 1$$

For maximum selectivity, low temperatures are essential: under such conditions, simple alcohols, even allylic, are unaffected, and with unsaturated lactones as substrates, the double-bond position stays unchanged.

General procedure (5)

A stirred solution of the lactol (7 mmol) and TESH (10.5 mmol) in dichloromethane (25 ml) was cooled to −78 °C, and boron trifluoride etherate (7.7 mmol) was added dropwise. Stirring was continued at −78 °C until t.l.c. indicated the absence of starting lactol. Saturated sodium hydrogen carbonate solution (10 ml) was added, the cooling bath was removed, and the mixture was allowed to come to ambient temperature with vigorous stirring. Ether (100 ml) was added, and the solution was washed with saturated sodium hydrogen carbonate solution (20 ml) and brine (20 ml). It was then dried and concentrated, and the product was isolated by chromatography, yields 50–88%.

REFERENCES

1. *Reviews*: E. Lukevics, Z. V. Belyakova, M. G. Pomerantseva and M. G. Voronkov, *Organometal. Chem. Rev.* **5**, 1 (1977); J. L. Speier, *Adv. Organometal. Chem.* **17**, 407 (1979).
2. M. F. Semmelhack and R. N. Misra, *J. Org. Chem.* **47**, 2469 (1982).
3. I. Ojima and T. Kogure, *Organometallics* **1**, 1390 (1982).
4. *Reviews*: D. N. Kursanov, Z. N. Parnes, M. I. Kalinkin and N. M. Loim, *Ionic Hydrogenation and Related Reactions*. Harwood Academic Publishers, Chur (1985); D. N. Kursanov, Z. N. Parnes and N. M. Loim, *Synthesis* 633 (1974).
5. G. A. Kraus, K. A. Frazier, B. D. Roth, M. J. Taschner and K. Neuenschwander, *J. Org. Chem.* **46**, 2417 (1981). For applications to complex substrates, see for example G. A. Kraus and K. A. Frazier, *J. Org. Chem.* **45**, 4820 (1980); E. W. Colvin and I. G. Thom, *Tetrahedron* **42**, 3137 (1986).
6. E. Winterfeldt, *Synthesis* 617 (1975).

– 18 –

Organic Syntheses

18.1. ALKYNYL/VINYLSILANES

18.1.1. Preparation

(E)-3-Trimethylsilylprop-2-en-1-ol

$$HC\equiv CCH_2OH \longrightarrow Me_3SiC\equiv CCH_2OH$$

$$Me_3SiC\equiv CCH_2OH \longrightarrow (E)\text{-}Me_3SiCH\!=\!CHCH_2OH$$

Organic Syntheses **64**, 182 (1986).

3-Trimethylsilylbut-3-en-2-one

$$CH_2\!=\!CHBr \longrightarrow CH_2\!=\!CHSiMe_3 \longrightarrow CH_2\!=\!C(Br)SiMe_3$$

$$\longrightarrow CH_2\!=\!C(SiMe_3)CHOHCH_3 \longrightarrow CH_2\!=\!C(SiMe_3)\underset{\underset{O}{\|}}{C}CH_3$$

Organic Syntheses **58**, 152 (1978).

18.1.2. Reactions

Use of 3-trimethylsilylbut-3-en-2-one in conjugate addition/annelation reaction with 2-methylcyclohexenone after conjugate addition with lithium dimethylcopper.
Organic Syntheses **58**, 158 (1978); see also R. K. Boeckman, *Tetrahedron* **39**, 925 (1983).

18.2. ALLYLSILANES

18.2.1. Preparation

Silylation of 2-methylpropen-1-ol → 2-trimethylsilylmethylpropen-1-ol (a synthon/precursor for trimethylenemethane):

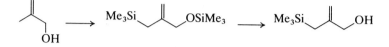

Organic Syntheses **62**, 58 (1984); the yield in this reaction can be improved to 98% by *not* removing the hexane from the n-BuLi, and omitting to add THF—B. M. Trost, S. M. Mignani and T. N. Nanninga, *J. Am. Chem. Soc.* **108**, 6051 (1986).

18.2.2. Reactions

$TiCl_4$-induced conjugate addition of allyltrimethylsilane to 4-phenylbut-3-en-2-one:

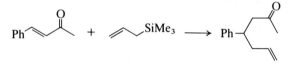

Organic Syntheses **62**, 86 (1984).

18.3. TRIMETHYLSILYL ETHERS

(i) Use of 2-methyl-2-trimethylsilyloxypentan-3-one (detailed preparation given) in diastereoselective aldol condensations, when it acts as a synthon for propanoic acid α-anion, after cleavage with HIO_4:

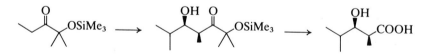

Organic Syntheses **63**, 79, 89 (1985).

(ii) Preparation of 1-ethoxy-1-trimethylsilyloxycyclopropane (and thence the ethyl hemiacetal of cyclopropanone) by reaction of ethyl 2-chloro-propanoate with Na/TMSCl:

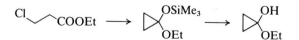

Organic Syntheses **63**, 147 (1985).

18.4. SILYL ENOL ETHERS

18.4.1. Preparation

(Z)-3-Trimethylsilyloxypent-2-ene

$$\underset{\underset{O}{\parallel}}{EtCEt} + Me_3SiCH_2COOEt \xrightarrow{TBAF} (Z)\text{-}Et\underset{\underset{OSiMe_3}{|}}{C}{=}CHCH_3$$

Use of ethyl trimethylsilylacetate (detailed preparation given) and TBAF (drying details: Fluka trihydrate dried over P_2O_5/48 h/0.1 mmHg, and then powdered in a dry atmosphere) for the preparation of (Z)-3-trimethylsilyloxypent-2-ene from pentan-3-one.

Organic Syntheses **61**, 122 (1983); non-enolizable aldehydes react differently, giving Reformatsky products (E. Nakamura, M. Shimuzu and I. Kuwajima, *Tetrahedron Lett.* 1699 (1976)).

1-Methoxy-3-trimethylsilyloxybuta-1,3-diene

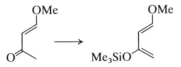

Preparation of 1-methoxy-3-trimethylsilyloxybuta-1,3-diene (Danishefsky's diene) and Diels–Alder addition with maleic anhydride.
Organic Syntheses **61**, 147 (1983).

2-Trimethylsilyloxybuta-1,3-diene

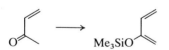

Preparation of 2-trimethylsilyloxybuta-1,3-diene and Diels–Alder addition with diethyl fumarate.
Organic Syntheses **58**, 163 (1978).

1,2-Bis(trimethylsilyloxy)cyclobutene

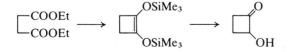

Preparation of 1,2-bis(trimethylsilyloxy)cyclobutene from diethyl succinate (Rühlmann condensation using Na and TMSCl), and of 2-hydroxycyclobutanone.
Organic Syntheses **57**, 1 (1977).

18.4.2. Reactions

t-Alkylation, using the reaction between 1-trimethylsilyloxycyclopent-1-ene (detailed preparation given) and 2-chloro-2-methylbutane, induced by TiCl$_4$.

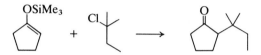

Organic Syntheses **62**, 95 (1984).

Cyclohept-2-enone

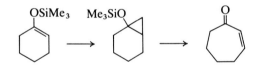

Ring expansion of trimethylsilyloxycyclopropanes, using the reaction between 1-trimethylsilyloxycyclohex-1-ene and CH$_2$I$_2$/Et$_2$Zn. FeCl$_3$-catalysed opening gives 3-chlorocycloheptanone, and thence cyclohept-2-enone.
Organic Syntheses **59**, 113 (1979).

Cyclobutane-1,2-dione

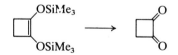

Oxidative cleavage of 1,2-bis(trimethylsilyloxy)cyclobutene with Br$_2$, to give cyclobutane-1,2-dione.
Organic Syntheses **60**, 18 (1981).

6-Hydroxy-3,5,5-trimethylcyclohexenone

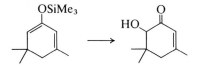

Oxidation with mcpba→ α-hydroxyketones, using the reaction of isophorone dienol ether (detailed preparation given), and $Et_3N.HF$ cleavage of the intermediate silyl ether.
Organic Syntheses **64**, 118 (1986).

2-Acetonylcyclohexanone

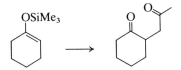

Conjugate addition to unsaturated nitro-compounds, using the reaction between 1-trimethylsilyloxycyclohex-1-ene and 2-nitropropene, under the influence of $SnCl_4$, to give 2-acetonylcyclohexanone (i.e. the nitro-compound acts as an umpolung reagent for an acetone moiety).
Organic Syntheses **60**, 117 (1981).

18.5. TMSX REAGENTS

18.5.1. Preparation/Reactions

TMSCN

Detailed preparation from LiCN (from LiH and acetone cyanohydrin) and TMSCl, followed by its reaction with *p*-benzoquinone in the presence of 18-Crown-6 and KCN.
Organic Syntheses **60**, 126 (1981).

TMSCN

(i) *In situ* preparation using $TMSCl/KCN/Zn(CN)_2$, and reaction with *p*-methoxybenzaldehyde.
Organic Syntheses **62**, 196 (1984).

(ii) Preparation of benzophenone cyanohydrin, by reaction of benzophenone with TMSCN/ZnI$_2$, and then HCl hydrolysis of the cyanohydrin trimethylsilyl ether.
Organic Syntheses **60**, 14 (1981).

TMSCN

Conversion of epoxides into β-hydroxy isocyanides—preparation of *trans*-2-isocyanocyclohexanol, using TMSCN to open cyclohexene oxide with *trans* stereochemistry, followed by KF/MeOH cleavage of the intermediate silyl ether.
Organic Syntheses **64**, 39 (1986).

TMSI

Preparation from hexamethyldisiloxane and I$_2$/Al powder in detail, followed by cleavage of cyclohexyl methyl ether, to give cyclohexanol (*via* the intermediate silyl ether).
Organic Syntheses **59**, 35 (1979).

Tris(dimethylamino)sulphonium difluorotrimethylsilicate

$$3Me_3SiNMe_2 \; + \; SF_4 \; \xrightarrow[25\,°C]{Et_2O} \; (Me_2N)_3S^+F_2SiMe_3^- \; + \; 2Me_3SiF$$

Organic Syntheses **64**, 221 (1986).

18.6. IONIC HYDROGENATION

(i) Et$_3$SiH/BF$_3$.Et$_2$O reduction/deoxygenation of *m*-nitroacetophenone, to give ethyl *m*-nitrobenzene.
Organic Syntheses **60**, 117 (1981).

(ii) Cl$_3$SiH reduction/deoxygenation of ArCOOH → ArCH$_3$.
Organic Syntheses **56**, 83 (1977).

– 19 –

Organic Reactions

The acyloin condensation
J. J. Bloomfield, D. C. Owsley and J. M. Nelke, *Organic Reactions* **23**, 259 (1976).
Note. Includes Rühlmann condensation.

The directed aldol reaction
T. Mukaiyama, *Organic Reactions* **28**, 203 (1982).
Notes. TMS enol ethers with aldehydes, ketones, using $TiCl_4$; acetals using TMSOTf. β-Ketosilanes with aldehydes.

Addition and substitution reactions of nitrile-stabilized carbanions
S. Arseniyadis, K. S. Kyler and D. S. Watt, *Organic Reactions* **31**, 1 (1984).
Note. Includes $ArC^-(OTMS)CN$, and $HetAr^-(OTMS)CN$.

Syntheses using alkyne-derived alkenyl- and alkynylaluminium compounds
G. Zweifel and J. A. Miller, *Organic Reactions* **32**, 375 (1984).
Note. Includes preparation of vinylsilanes and α-halogenovinylsilanes.

– 20 –

Organometallic Syntheses

J. J. Eisch, *Organometallic Syntheses*, Vol. 2, ed. J. J. Eisch and R. B. King. Academic Press, New York (1981).

p.159: *Phenyl trimethylsilylethyne*

$$PhC\equiv CH \xrightarrow{EtMgBr} PhC\equiv CMgBr \xrightarrow{TMSCl} PhC\equiv CSiMe_3$$

p. 160: *(Z)- and (E)-β-styryltrimethylsilane*

$$(E)\text{-PhCH}{=}\text{CHSiMe}_3 \xleftarrow[2.\ H_2O]{1.\ R_2AlH} PhC\equiv CSiMe_3 \xrightarrow[2.\ H_2O]{1.\ R_2AlH.R_3N} (Z)\text{-PhCH}{=}\text{CHSiMe}_3$$

p.161: *Triphenylsilyl lithium and -potassium*

Organometallic Syntheses, Vol. 3, ed. R. B. King and J. J. Eisch. Elsevier, Amsterdam (1986).

p.492: *(E)-1,2-Bis(trimethylsilyl)ethene* (J.-P. Pillot, P. Lapouyade and J. P. Dunoguès)

$$Me_3SiCH{=}CH_2 + TMSCl + Mg \xrightarrow{HMPA} (E)\text{-Me}_3SiCH{=}CHSiMe_3$$

p.494: *1,3-Bis(trimethylsilyl)propyne* (B. J. M. Bennetau and J. P. Dunoguès)

$$(Me_3Si)_2C{=}C{=}C(SiMe_3)_2 + TFA \xrightarrow[0\,^\circ C]{CCl_4} Me_3SiC\equiv CCH_2SiMe_3$$

139

p. 514: *Optically-active allyl silanes* (T. Hayashi and M. Kumada)

$$PhCHMgBr + CH_2=CHBr \xrightarrow{PdCl_2[(R)-(S)-PPFA]} CH_2=CH\overset{SiMe_3}{\underset{H}{C}}*Ph$$
$$\underset{SiMe_3}{|}$$

p. 524: *Potassium organopentafluorosilicates* (K. Tamao)

$$n\text{-}C_8H_{17}SiCl_3 + 5KF \longrightarrow K_2[n\text{-}C_8H_{17}SiF_5] + 3KCl$$

p. 527: *Tetrakis(trimethylsilyl)allene* (B. J. M. Bennetau, D. Y. N'Gabe and J. P. Dunoguès)

$$C_6Cl_6 + Li + TMSCl \xrightarrow[0\,°C]{THF} (Me_3Si)_2C=C=C(SiMe_3)_2$$

p. 536: *2-Trimethylsilylpyridine* (M. Behrooz and J. J. Eisch)

p. 544: *Geranyltrimethylsilane* (E. Negishi and F.-T. Luo)

$$\text{geranyl chloride} \xrightarrow[THF]{LiTMS} \text{geranyltrimethylsilane}$$

Index of Compounds and Methods